An Introduction to Engine Testing and Development

Other SAE titles of interest:

Engine Testing: Theory and Practice, Third Edition
By A.J. Martyr and M.A. Plint
(Product Code: R-382)

Modern Engine Technology from A to Z
By Richard van Basshuysen and Fred Schäfer
(Product Code: R-373)

**Internal Combustion Engine Handbook:
Basics, Components, Systems, and Perspectives**
By Richard van Basshuysen and Fred Schäfer
(Product Code: R-345)

Introduction to Internal Combustion Engines, Third Edition
By Richard Stone
(Product Code: R-278)

Advanced Engine Technology
By Heinz Heisler
(Product Code: R-163)

For more information or to order a book, contact SAE International at
400 Commonwealth Drive, Warrendale, PA 15096-0001;
phone (724) 776-4970; fax (724) 776-0790;
e-mail CustomerService@sae.org;
website http://store.sae.org.

An Introduction to Engine Testing and Development

Richard D. Atkins

Warrendale, Pa.

All rights reserved. No part of this publication may be reproduced, stored in a retrieval system, or transmitted, in any form or by any means, electronic, mechanical, photocopying, recording, or otherwise, without the prior written permission of SAE.

For permission and licensing requests, contact:

SAE Permissions
400 Commonwealth Drive
Warrendale, PA 15096-0001 USA
E-mail: permissions@sae.org
Tel: 724-772-4028
Fax: 724-772-4891

Library of Congress Cataloging-in-Publication Data

Atkins, Richard D., 1940-
  An introduction to engine testing and development / Richard D. Atkins.
     p. cm.
  Includes index.
  ISBN 978-0-7680-2099-1
  1. Automobiles--Motors--Testing. 2. Automobiles--Motors--Design and construction. I. Title.

  TL210.A755 2009
  629.25--dc22   2008049298

SAE International
400 Commonwealth Drive
Warrendale, PA 15096-0001 USA
E-mail: CustomerService@sae.org
Tel: 877-606-7323 (inside USA and Canada)
            724-776-4970 (outside USA)
Fax: 724-776-1615

Copyright © 2009   SAE International

ISBN 978-0-7680-2099-1

SAE Order No. R-344

Printed in the United States of America.

Contents

Introduction xiii

Chapter 1   **The Test Facility and Methods of Measuring Engine Power** 1
- Test Facilities and Test Cells 1
- Dynamometers 4
  - Function of the Dynamometer 7
  - How the Dynamometer Works 7
  - Dynamometer Types and Operating Principles 7
  - Operating the Dynamometer 11
  - Hydraulic Water Brakes 11
  - Electric Dynamometers 12
  - Mechanism of the Dynamometer 15
  - Dynamometer Characteristics 17
  - Selection of Prop Shafts 19

Chapter 2   **In-Cell Services** 21
- Raw Water Services (Cooling) 21
  - Function of the Raw Water Cooling Circuit 22
  - Site Water Services 22
- Air Services 29
  - Combustion or Induction Air 29
  - Air Cooling and Ventilation 32
- Docking Rigs 33
- Some Engine Testing Pointers 36
  - Pre-Start, Operating the Dynamometer 36
  - Predictive Analysis 36
  - A Transient Test 37
- The Key to Control Systems 37
  - Proportional Response 39
  - Integral Response 39
  - Derivative Response 39
  - Tuning 40

Chapter 3   **Instrumentation: Temperature, Pressure, Flow, and Calibration** ... 41
- Temperature: The Principle and Application of Thermocouples 41
  - The Principle of Thermocouple Operation 41
  - The Law of Interior Temperatures 42
  - Standardized Thermocouples and Categories 44
- Pressure: A Review of Pressure Measuring Devices 44
  - What is Pressure? 44
  - Pressure Measuring Devices 45
  - Transducers 48
  - Types of Pressure Transducers 51
- Flow Measurement 52
  - Mass Airflow Sensors 53
  - Rotameters 54
  - Square Edged Orifice Plates 54
  - Lucas-Dawe Air Mass Flow Meters 56

		Calibration 56
		Definitions of Calibration Terms...............................57
		Calibration Personnel and Equipment.......................58
		Temperature Calibration..58
Chapter 4		**An Introduction to Mr. Diesel and His Engines**..........................61
		Unit Injectors 65
		Advantages and Shortcomings of Diesel Engines......................65
		Disadvantages of Diesel Engines................................65
		Advantages of Diesel Engines 66
		Why Does the Diesel Engine Have Such Good Fuel Consumption?..67
		Fuel Injection and Combustion Principles..................................68
		Pre-Chamber Systems...68
		Direct Injection Systems 70
		Comparison of the Systems..72
		An Introduction to Diesel Fuels..72
		Cetane Number..72
		Cold Behavior 72
		Density 72
		Calorific Value..72
		The Diesel Engine and International Regulations 73
		Diesel Particulate Filters...73
		Continuously Regenerating Trap Filters.....................................75
Chapter 5		**Engine Tests Used Within the Automotive Testing Industry** 77
		Types of Tests 77
		Understanding Durability Testing...78
		Definitions...79
		Reliability..80
		Durability 80
		In-Cell Testing 81
		Early Development Phase Tests..82
		Defining the In-Cell Test Procedure 82
		Increasing the Severity of the Test...82
		Thermal Stress 84
		Thermal Shock Testing 84
		Combining Bench Testing with In-Field or Trials Testing 84
		Test Duration and Engine Life Comparison 85
		Pass/Fail Checklist..86
		Summary..86
		Example of a Typical Test Procedure for a High-Speed Diesel Engine 250-Hour Validation Test 86
		Master Service Record Sheet..94
		Interpretation of the Test Results 95
		Total Preventive Maintenance: Daily Test Bed/Cell Checks..........95
		Recordkeeping...96
		Reasons for Keeping Records 96
		Types of Records...97
		Sales and Marketing 98
		Fault Diagnosis 98
		Reporting 99

Chapter 6	**Spark Plugs**..101	
	Spark Plug Ratings 103	
	Running a Spark Plug Rating Test..104	
	Radio Frequency Interference..108	
	Summary..108	
Chapter 7	**Exhaust Gas Emissions and Analysis**...111	
	Exhaust Gas Emissions..111	
	Group A—Gases that Can Cause Death or Injury	
	Within Minutes..111	
	Groups B and C—Gases that Can Cause Death or	
	Serious Illness with Prolonged Exposure and Can	
	Create Minor Health Problems or Are a Nuisance.................112	
	Group D—Gases Associated with Global Warming 113	
	Simple Combustion Theory, Ideal Combustion, and	
	Stoichiometry..113	
	Of What Is Fuel Composed? 113	
	What Are Diesel Emissions?..114	
	What Makes Up Air?...122	
	How Does Combustion Occur?..122	
	Why Do Internal Combustion Engines Emit Gases? 125	
	Balanced Chemical Equations...127	
	Non-Ideal Combustion—Formation of Pollutants 129	
	Fuel Droplets..129	
	Combustion Differences Between Diesel and Gasoline Engines ...130	
	Direct and Indirect Injection Diesel Engines................................131	
	Operating Principles of Emission Reduction Devices Fitted	
	to Internal Combustion Engines 137	
	Catalyst Operation..137	
	Exhaust Gas Recirculation 138	
	Effects of Brake Mean Effective Pressure 138	
	Principles and Operation of Raw, Dilute, Continuous,	
	and Bag Sampling...138	
	Dilute and Raw Sampling 138	
	Particulate Measurement..139	
	Principles and Operation of Flow Tunnels...............................139	
	Filter Handling and Weighing 142	
	Principles and Operation of Micro-Tunnels...........................142	
	Principles and Operation of Mini-Tunnels.............................143	
	Principle of Continuous Particulate Analyzers 143	
	Hot and Cold Sampling..144	
	Essential Elements of a Sampling System...................................144	
	Temperature Control 144	
	Pressure Control 144	
	Flow Rate 145	
	Dryness...145	
	Solid Particles...145	
	Sample Pump 145	
	Leak Checking 145	
	Back Flushing...145	
	Non-Dispersive Infrared Analyzer 145	
	Flame Ionization Detector..147	

NOx Analyzers 148
Oxygen Analyzers 148
Time Alignment 150
Maintenance 151
Calibration of Analyzers Using a Gas Divider—
 NOx Efficiency Checks 151
Operation of Smoke Meters 151
Operation of Smoke Meters Using the Filter Paper
 Technique 153
Operation of Smoke Meters Using the Opacity Technique 153
Correct Venting of Emission Analyzers 155
Factors in Monitoring Exhaust Gases 155
Compounds Present in Vehicle Exhaust Gases 156
Emission Testing 156
New Exhaust Gas and Particulate Measuring Methods 158
Laser 158
Combustion Fast Response Analyzers for Transient Work 158
Summary 158
After-Treatment Considerations 158
Exhaust Gas Recirculation 158
Development of Low Emission Fuels 159

Chapter 8 **Combustion Analysis** 161
Basic Combustion 161
Internal Combustion Engine 162
Intake Stroke 162
Compression Stroke 162
Ignition Stroke 163
Power Stroke 164
Exhaust Stroke 165
Cylinder Pressure Measurements 166
Efficiency Loss Mechanisms in the Vehicle Drivetrain 168
What Are the Efficiency Loss Mechanisms? 168
Heat Losses 171
Efficiency Overview 175
Features of Combustion Analysis and Diagnostics 175
Brake Mean Effective Pressure 177
Indicated Work—Indicated Mean Effective Pressure 177
Pumping Mean Effective Pressure 178
Net Mean Effective Pressure 178
Frictional Mean Effective Pressure and Mechanical
 Efficiency 179
Indicated Efficiency 179
Volumetric Efficiency 179
Types of Combustion Diagnostics 180
Non-Heat Release 180
Heat Release 182
Types of Heat Release Algorithms 182
Thermodynamic Heat Release 184
Recommendations 185
Combustion Variability 186
How Does Combustion Variability Manifest Itself? 186

	Causes	186
	Impact	186
	How Is Combustion Variability Quantified?	188
	Abnormal Combustion	194
	Incomplete Combustion	194
	Causes of Misfire	194
	Causes of Partial Burns	195
	Knock	196
	Pre-Ignition	199
	Calibration Issues	200
	Encoders	211
	Data Integrity	212
	Control Charts	213
	Good Test Practices	215
	Recommended Daily Checks	216
Summary		217
Some Calculations Used in Conjunction with Combustion Analysis Work		217
	Combustion Efficiency	217
	Determining the Effective Compression Ratio	218

Chapter 9 Turbochargers 221
Introduction 221
Pulse Energy 221
Charge Air Cooling 222
Turbocharger Matching 225

Chapter 10 Fitting Operations Within the Test Cell Best Practices 233
Stripping 233
Checking and Inspecting 233
Rebuilding 234
Golden Rule for Handing Over 234
Cleaning 234
Quality 234
Commonly Occurring Incidents When Testing Engines on Dynamometers 234
    Damaged Valve or Piston and Bore 234
    Timing Belt "Jumping a Tooth or Three" 235
    Sheared or Snapped Prop Shaft 235
Summary 235

Chapter 11 The Basic Internal Combustion Engine 237
Introduction 237
Engine Lubrication System 238
    Lubrication System Components 238
    Instrumentation of the Lubrication System 241
Engine Cooling System 241
    Cooling System Components 241
    Instrumentation of the Cooling System 244
Engine Induction System 245
    Induction System Components 245
    Instrumentation of the Inlet Manifold 247

	Engine Exhaust System 248
	Exhaust Manifold System 248
	Instrumentation of the Exhaust System.....................................248
Chapter 12	**Quality Standards for Engine Testing**..249
	Quality Standards for Test Laboratories 250
	BS EN ISO 9000 Series 250
	United Kingdom Accreditation Service (UKAS) (EN 45001)...250
	Implications of Quality Standards 251
Chapter 13	**Base Calculations** 253
	Torque Backup...253
	Motoring Mean Effective Pressure 253
	Volumetric Efficiency 253
	Specific Fuel Consumption 254
	Correction Factors 254
	Phase 255
	Cycle 255
	Process 255
	Heat..255
	Enthalpy 256
	Specific Enthalpy 256
	Principle of the Thermodynamic Engine 256
	Mechanical Power 256
	Electrical Power...257
	Laws of Thermodynamics 257
	The Conservation of Energy 257
	Joule's Law 257
	Entropy 258
	Correction Formulae..260
	Examples of Calculations Required Within the Test Cell Environment 262
	Test Bed Fuel Flow Measurement..262
	Test Bed Airflow Measurement..262
	Brake Specific Fuel Consumption..262
	Brake Specific Air Consumption..263
	Efficiencies 264
	Thermal Efficiency...264
	Mechanical Efficiency..264
	Volumetric Efficiency...265
	Air/Fuel Ratio 265
	First Law of Thermodynamics..266
	Second Law of Thermodynamics 266
	Mathematical Basis for Power Correction Factor 268
	Swept Volume...270
	Compression Ratio 270
	Brake Mean Effective Pressure 270
	Pumping Mean Effective Pressure 270
	Compression/Expansion Mean Effective Pressure 271

  Friction Mean Effective Pressure 271
  Brake Torque and Power 271
  Motoring Mean Effective Pressure 272

Glossary of Terms and Acronyms Used in the Design and Testing of Internal Combustion Engines 273

Index 283

About the Author 291

Introduction

How does one describe the internal combustion engine? My grandfather, a renowned mechanical engineer, defined it for me as "the infernal confusion engine." Now, many years later, with countless research and development projects behind me, I believe that he was possibly more accurate than he ever envisaged.

Simply put, the internal combustion engine is an energy conversion device that converts thermal energy (heat) into mechanical energy. When hydrocarbon fuel is burned in air, some of the chemical energy contained in the fuel is converted into work. The nitrogen trapped in the cylinder is heated by the energy released when the carbon and hydrogen in the fuel react with oxygen in the air. That's all there is to it.

It is generally accepted that a Dutch physicist, Christian Huygens, first formulated the principle of the internal combustion engine in 1680; however, he proposed to use gunpowder as the motive power source. In 1860, Etienne Lenoir, a self-taught mechanic, revealed to the world the first internal combustion engine that worked. For some time, engineers had understood both the low efficiency of the steam engine and the desirability of a device that would burn its fuel inside the cylinder instead of using it to produce steam as an intermediary. Many engines were designed, but Lenoir's engine was the first recorded engine to pass the experimental stage. It was fueled by coal gas (used at that time primarily for street and home lighting). This coal gas was mixed with air and was drawn into a cylinder by withdrawal of a piston. At the midpoint, electric sparks ignited the mixture, so that only the second half of each stroke was powered. However, Lenoir's engine was double acting, so fuel entered either side of the piston in turn. The operating principle closely followed that of the steam engine. The engine ran, but it was very inefficient, the gas was expensive, and vast quantities of gas were used (almost 3 cubic meters) to produce 0.75 kW at 100 rev/min. In addition, it had a significant noise and vibration problem, producing violent shocks at each explosion, which Lenoir attempted to damp out by the use of springs and other devices to absorb the power stroke shocks. In 1862, a railway engineer named Alphonse Eugène Beau de Rochas published a pamphlet about improvement in locomotive design, in which he suggested compounding the steam engine with gas engines. This was a significant advance, and to date, its principles have not been challenged. He stated that the gas in the engine should ignite continuously under high compression, which was instigated by making it work in four stages:

1. Stage one—Intake during one whole stroke of the piston

2. Stage two—Compression during the following stroke

3. Stage three—Firing at the dead point, and expansion during the third stroke

4. Stage four—Expulsion of the burned gases from the cylinder

Thus, Beau de Rochas presented the principle of the four-stroke engine as we understand it today. However, he never undertook to construct an engine on these lines nor to present a paper to his peers. As frequently happens in life, quite independently of Beau de Rochas, Nikolaus August Otto, a German traveling salesman who was fascinated by all things technical and was blessed with an inquiring mind, took up the invention. The problem of enabling the engine to run efficiently by controlling the richness of the gas/air mixture presented a great challenge. Otto produced an engine in which the expulsion drove the piston upward into a vertical cylinder, where the contraction of the spent fuel as it cooled produced a vacuum, into which atmospheric pressure and gravity forced the piston back. This followed the system used by the steam-driven beam engines of the eighteenth century. The engine worked, and while it was better than the Lenoir design, it still was a vibrating, noisy, and inefficient device.

Tradition has it that one day in 1875, Otto was watching a smoking chimney, and his imagination was caught by the smoke that first emerged in dense plumes and then gradually dissipated into the air. His supposition was that it should be possible to introduce a rich fuel/air mixture to the point of ignition, where it would be cushioned from the piston by a much thinner layer of inert air next to it. This principle of stratification was almost certainly, in this instance, incorrect. However, to produce it, Otto reinvented the four-stroke cycle as Beau de Rochas had envisaged it (but henceforth was called the "Otto cycle") and embodied it in his "silent Otto engine" of 1876, which was a tremendous success, 2 kW at 180 rev/min. The engine utilized a very dangerous open flame ignition system to fire the coal gas fuel that Otto was using. (Lenoir used an advanced spark ignition system.) In parallel with Otto's work, an Austrian inventor named Siegfried Marcus in 1867 had invented a carburetor to convert liquid petroleum into flammable gas.

In 1861, Otto patented a two-stroke engine that ran on gas. Otto and his partner, a German industrialist named Eugen Langen, built a factory and worked on improving the engine. Their two-stroke engine won a gold medal at the 1867 World's Fair in Paris. The company was named N.A. Otto & Co., which was the first company to manufacture internal combustion engines. The company exists today as Klockner-Humbolt-Deutz AG, the oldest company manufacturing internal combustion engines and the world's largest manufacturer of air-cooled diesel engines.

In May 1876, Otto built the first four-stroke piston cycle internal combustion engine. This was the earliest practical alternative to the steam engine. In the next ten years, more than 30,000 of the engine were sold. This engine was the prototype of all combustion engines that have since been built. The operating principle of the engine was named the "Otto cycle" in honor of Nikolaus Otto. The design of the engine consists of four stokes of a piston, which draw in and compress a gas/air mixture within a cylinder. This process results in an internal explosion. Otto's gas/motor engine had the patent number 365,701. In 1862, Alphonse Beau de Rochas, a French engineer, had patented the four-stroke cycle. However, Otto was the first to build a four-stroke cycle engine. Nevertheless, in 1886, Otto's patent was revoked when Beau de Rochas' patent was revealed. Nikolaus August Otto died on January 26, 1891.

Meanwhile, Gottlieb Daimler constructed a very light engine using Otto's model and attached one of them to a bicycle. This became the world's first motorcycle. Karl Benz built his first three-wheeled automobile employing Otto's engine. Daimler also constructed an automobile using Otto's engine. The firms of Daimler and Benz merged and manufactured the famous Mercedes-Benz vehicles. George Brayton, an American engineer, developed a two-stroke kerosene engine in 1873, but it was too large and too slow to be commercially successful.

In 1885, Gottlieb Daimler constructed what generally is recognized as the first modern high-speed internal combustion engine. Small and fast with a vertical cylinder, the engine used gasoline inducted via a carburetor. In 1889, Daimler introduced a four-stroke engine with mushroom-shaped valves and two cylinders arranged in a "V" configuration, having a much higher power-to-weight ratio. With the exception of electric starting, which would not be introduced until 1924, all modern gasoline engines are descended from Daimler's engines.

The internal combustion engine as we understand it has been with us now for some 120 years. With the advent of computer-aided design systems, flow visualization, and highly advanced mathematical models, one would think it would be possible to design and manufacture the ideal engine the first time, every time. If only this were the case.

There is no doubt that fantastic advances are being made at an accelerating rate, but the internal combustion engine is an extremely complex device, requiring a sound understanding of many disciplines. With ever diminishing world oil stocks and increasingly stringent government legislation worldwide, the challenges facing the automotive engineer have never been greater.

Downsizing, the intelligent use of new technologies, and incorporation of these into cost-effective vehicles will present the automotive engineer with stimulating and challenging work for decades to come. My purpose in writing this book is to present some of the basic principles required in the testing and development of the internal combustion engine powertrain system, thus giving the new automotive engineer the basic tools required to meet these challenges.

Richard D. Atkins
Hastings, 2009

Chapter 1

The Test Facility and Methods of Measuring Engine Power

Test Facilities and Test Cells

To test an engine, we must be able to compare the performance of differing states of tune and different types of engines. To do this, a complex control and data acquisition system is required. Much detailed planning is needed before setting up such a facility. Requirements for complex test procedures to satisfy ever demanding regulations have elevated the status of an engine or vehicle test cell into that of a sophisticated laboratory. Such a laboratory is heavily dependent on advanced testing equipment and instruments. This apparatus requires equally sophisticated and, more importantly, reliable support services and utilities to maintain its reliability and to deliver the performances expected. Similarly, new and complicated test procedures require the strict control of various parameters such as temperature, pressure, flow, humidity, velocities, and so forth, all of which lie within narrow tolerance bands. Furthermore, test laboratories are required to comply with international and local legislation standards and governmental requirements such as the Health and Safety at Work Regulations (1999), Control of Substances Hazardous to Health (COSHH), building regulations, petroleum and fire officer requirements, ISO standards, and BS/BS EN standards (these being generic names given to a family of standards developed to provide a framework around which a quality management system can effectively be established).

Regardless of the purpose of the test to be conducted within a cell running any form of a test—be it a simple durability test or a complicated analytical test—the operation is an expensive exercise and thus must be accomplished correctly the first time with no failures or any form of interruption. Repeating a test unnecessarily even once due to any form of failure or interruption is expensive, time consuming, and, in some instances, could result in the loss of business due to unreliability.

The services that need to be installed in a test facility include such items as ventilation; treated air; cooling water; and conditioning systems for fuel, lubricant, coolant, and combustion air. Other items that also must be considered include compressed air; exhaust gas disposal; analytical, span, and other reference gases; lighting and emergency lighting; power and small power; control supplies; and detection, alarm, and suppression systems. In reality, the utilities are not limited to only those mentioned here but include the provision of acoustics, noise, and vibration measuring and monitoring equipment; fuel systems storage and distribution; gas systems storage; and utilities for the control rooms, operator areas, workshops, laboratories, offices, and so forth. It is truly a task for the professional.

The facility does not end with the test cell. A clean engine build area is required, as well as a pre-installation rig area, an instrumentation store, a fuel farm, a general parts store, a dirty strip and wash area, a study area for the engineers and technicians, and toilet and washing facilities. To ensure smooth operation of the facility, the total site must be planned in great detail. Figure 1.1 shows a specimen layout of such a facility, and Figure 1.2 shows a typical development facility.

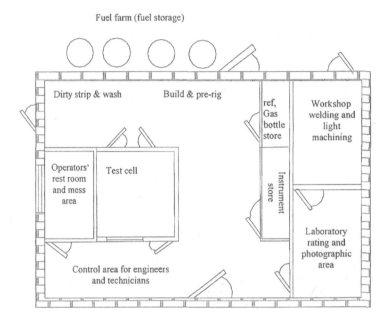

Figure 1.1 A typical test facility layout.

Figure 1.2 A typical development facility. (Courtesy of Froude Hoffman)

In designing the cell, the cooling plant and air exchange through the cell limit the size and power of the engines that can be tested. (Compared with the cost of the cooling plant and air exchange, the dynamometer is relatively inexpensive.) With this in view, the recommended action is to attempt to estimate the maximum power output that will need to be tested over the next 10 years and then add 50% to the capacity of the cooling and air-exchange systems. These systems can simply be throttled down, if required, to suit any specific application.

When setting out the specification of a new test facility, it is recommended that a series of energy balance calculations for a number of applications be undertaken. This work entails noting the fuel rate of the engine at maximum rated speed and load, taking into account the relative efficiency of the engine (for example, 26% for a typical gasoline application and 34% for a typical diesel-powered engine), then identifying where the remaining heat energy is lost: heat to exhaust, heat to coolant, heat to oil, radiated heat from the engine, and so forth. From this, one can review the amount of air circulation required within the cell. The air that the engine inducts is only a small part of the whole. A great deal of radiated heat must be dissipated, and this requires a significant airflow—up to five complete cell air changes per minute.

Referring to Figure 1.3, the energy balance of a 2.4-liter EURO 3 engine manufactured by the Andoria engine company in Poland is reviewed. This power unit produced 75 kW at the flywheel at a rated speed of 4100 rev/min. The specific fuel consumption at net peak torque (ISO 1585) was 255 g/kWh.

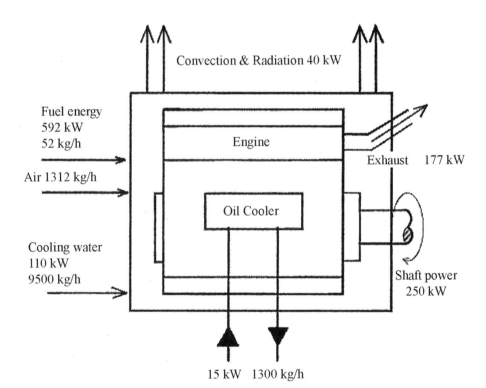

Figure 1.3 Energy into the engine; energy out of the engine.

$$\text{Brake specific fuel consumption} = \frac{\text{Fuel used (g/H)}}{\text{Power (kW)}}$$

$$255 \text{ (BSFC)} = \frac{\text{Fuel used}}{75 \text{ kW}}$$

$$\therefore \quad \text{Fuel used (liters)} = 75 \times 0.255 = 19.125 \text{ kg/h}.$$

The fuel used with this specific fuel consumption calculates out at 19.125 kg/h. The specific gravity of this fuel was not known, but it would lie between 0.815 and 0.855 kg/liter. For the preceding calculations, the following were assumed:

$$\text{Fuel used} = 19.125 \times 0.815 = 15.58 \text{ liters}$$

$$3.8 \text{ liters of standard diesel} = 155 \times 10^6 \text{ joules} = 147,000 \text{ Btu}$$

$$1 \text{ kWh} = 3.6 \times 10^6 \text{ joules}$$

$$19.125 \text{ kg} = 15.58 \text{ liters} \times 40.79 \times 10^6 \text{ joules} = 176.53 \text{ kW}$$

Thus, the energy in is 176.53 kW, and the shaft energy out is 75 kW. (The remainder is shown in Figure 1.3.) Thus, it can be seen that this particular engine is 42.58% efficient (Figure 1.3).

The energy balance of the 75-kW turbocharged diesel engine described previously is shown in Table 1.1.

**TABLE 1.1
ENERGY BALANCE**

Item	Energy In	Item	Energy Out
Fuel	176.53 kW	Power	75 kW
		Heat to coolant	33 kW
		Heat to oil	4.5 kW
		Heat to exhaust	53.1 kW
		Convection and radiation	11 kW
Total	176.53 kW	Total	176.53 kW

When designing a test cell, consider all the heat-generating surfaces, such as those shown in Figure 1.4.

Dynamometers

William Froude (Figure 1.5) is regarded as the father of the modern dynamometer. His first project was to design a dynamometer for the steam engine in the *HMS Conquest* (Figures 1.6 through 1.9). The unit was fitted to the propeller shaft of the *HMS Conquest*, and the unit was submerged to provide cooling capacity for the absorbed power. Handles located on the stern of the ship operated a complex series of bevel gears that opened and closed sluice gates. An arrangement of levers read the torque on a spring balance located on the quay; a mechanical mechanism noted the engine revolutions. These were coupled to a rotating drum, and this produced a speed-versus-load chart, the area under the graph being the power.

The Test Facility and Methods of Measuring Engine Power

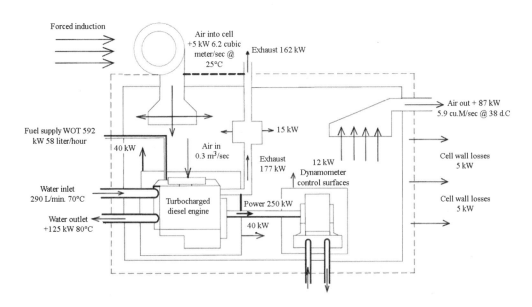

Figure 1.4 Typical test cell energy balance.

Figure 1.5 William Froude. (Courtesy of Froude Hoffman)

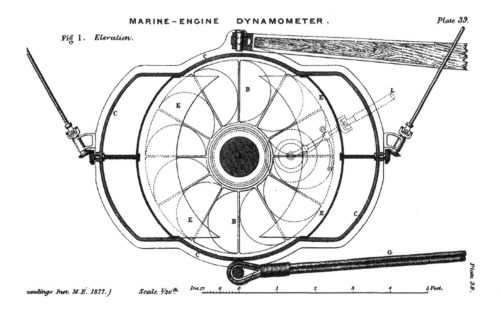

Figure 1.6 Original dynamometer drawing. (First published by the Institute of Mechanical Engineers in 1877)

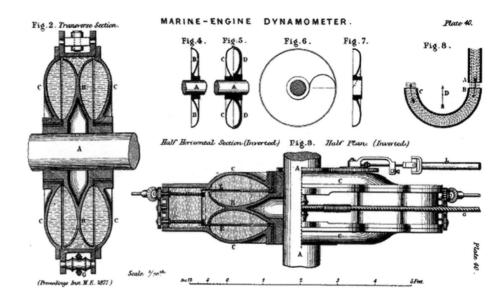

Figure 1.7 Original drawing, section of the rotor. (First published by the Institute of Mechanical Engineers in 1877)

Figure 1.8 Submerged original dynamometer. (Courtesy of Froude Hoffman)

The dynamometer is as fundamental to the in-cell testing of engines as is the engine. In establishing the engine characteristics and performance under different "road load" conditions, it is necessary to be able to safely and effectively replicate actual on-road

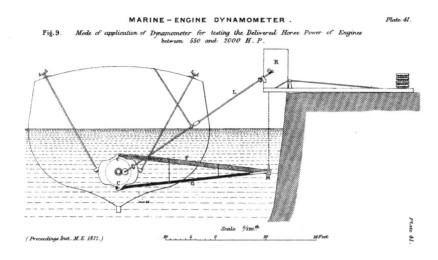

Figure 1.9 Original installation drawing. (First published by the Institute of Mechanical Engineers in 1877)

conditions on a consistent and repeatable basis. This is in essence what the dynamometer enables one to do when running engines are tested.

Function of the Dynamometer

The function of the dynamometer is to impose variable loading conditions on the engine under test, across the range of engine speeds and durations, thereby enabling the accurate measurement of the torque and power output of the engine.

How the Dynamometer Works

To better understand how the dynamometer works, imagine anchoring a spring balance to the ground, with a rope attached to the top eye and wrapped around a drum with a slipknot. The slipknot is tightened as the drum rotates, the rope then will be tensioned, and the balance will extend to indicate this tension as a weight (Figure 1.10).

Friction between the rope and the drum will slow the drum and its driving engine until, for example, at 1000 rev/min, the spring balance reads 210 kg. In effect, the weight being lifted is 210 kg, and the speed of the drum or engine then will be used in a formula to calculate the horsepower. In its simplest form, the engine is clamped on a test bed frame, and the output from the engine flywheel is connected to a driveshaft (propeller shaft) and hence a dynamometer.

Dynamometer Types and Operating Principles

Many types of dynamometers are available to the industry, with each having its own distinct advantages and disadvantages compared to those of its rivals. This section will focus on the main types found across the industry in general and will give only a brief outline of the other not-as-commonly-used variants. The main types of dynamometers considered here are as follows:

- "Hydraulic" (or water brake), of which there are two types:

  1. Constant fill
  2. Variable fill

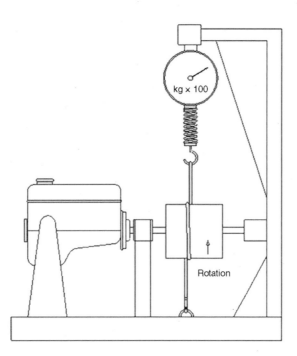

Figure 1.10 Friction rope dynamometer.

- Electrical, of which there are three main types:

  1. DC current
  2. AC current
  3. Eddy current

For many years, the design principle of the Froude dynamometer (Figures 1.6 to 1.9) was regarded as the industry standard dynamometer. While high-resolution transient systems are much favored today, an understanding of the design principles of these early hydraulic dynamometers is worthwhile. These dynamometers ranged in absorption values from 1 kW to more than 10,000 kW. In Figure 1.11, the rotor pockets and driveshaft of an ultra-large unit can be seen clearly. The water in the stationary pockets that is shearing those in the rotor pockets produces the torque reaction and the power absorption action.

Although the hydraulic dynamometer is based on a design that is more than 130 years old, it is still used in many applications, including the testing of Formula 1 engines and gas turbine aircraft engines.

The hydraulic water brake used in Formula 1 race engines will absorb more than 1000 BHP at speeds of greater than 20,000 rev/min (Figure 1.12).

The rotor is a critical design feature when running at very high speeds (greater than 20,000 rev/min). The example shown in Figure 1.13 has a diameter of 154 mm. This allows for a safety factor that is established by the bursting speed of the rotor. Unlike the grooves on either side of the rotor, these are labyrinth grooves and act as a noncontact seal.

From Figure 1.14, it can be seen that the main shaft is carried by bearings fixed in the casing. The casing in turn is carried by anti-friction trunnions, so that it is free to swivel about the same axis as the main shaft. When on test, the engine is coupled directly to the main shaft transmitting the power to a rotor revolving inside the casing. This

The Test Facility and Methods of Measuring Engine Power

Figure 1.11 Rotor pockets. (Courtesy of Froude Hoffman)

Figure 1.12 High-performance F-type water brake. (Courtesy of Froude Hoffman)

Figure 1.13 High-performance rotor. (Courtesy of Froude Hoffman)

Figure 1.14 Eddy current unit. (Courtesy of Froude Hoffman)

water is circulated to provide hydraulic resistance while simultaneously carrying away the heat developed by the destruction of power. In each face of the rotor are formed pockets of semi-elliptical cross sections divided from one another by means of oblique vanes (Figure 1.11).

The internal faces of the casing are provided with liners, which are pocketed in the same way. Thus, the pockets in rotors and liners together form elliptical receptacles around which the water moves at high speed. Where high torque absorption levels are required, twin rotor units are specified (Figure 1.14). When in action, the rotor discharges water at high speed from its periphery into pockets formed in the casing liners, by which it then is returned at diminished speed into the rotor pockets at a point near the shaft. The resistance offered by the water to the motion of the rotor reacts upon the casing, which tends to turn on its anti-friction roller supports. This tendency is counteracted by means of a lever arm terminating in a weighing device that measures the torque.

The forces resisting rotation of the dynamometer shaft may be divided into three main classes:

1. The hydraulic resistance created by the rotor
2. The friction of the shaft bearings, which usually are of the ball type
3. The friction of the sealing glands

Note that every one of these forces reacts upon the casing, which, being free to swivel upon anti-friction trunnions, transmits the whole of the forces to the weighing apparatus. "Every force resisting rotation of the engine shaft is caused to react upon the weighing apparatus."

The dynamometer shaft is of robust construction with very high torque absorption characteristics. These are required to cope with the demands of resisting the rotational force (or torque) of the engine power unit to which it is coupled by the prop shaft/driveshaft.

The dynamometer shaft is connected to the driveshaft (or prop shaft), which in turn is bolted to the engine crankshaft. This means that the engine crankshaft and the dynamometer shaft will be running at the same speed (revolutions per minute [rpm]). This again means that as the rotor is connected to the dynamometer shaft, the rotor is driven at the same speed as the engine crankshaft.

The quantity of water required to carry away the heat generated by the absorption of power can be calculated. Each kilowatt that is absorbed generates 14,338 calories per minute, most of which passes into the cooling water.

Operating the Dynamometer

Prior to starting the engine, open the inlet valve fully and the outlet valve very slightly. It is almost always advisable to start up with a light load, and this may be accomplished by screwing the sluice gates as far into the dynamometer as they will go. The engine is now started. To regulate the load, open the sluice gates by means of the handwheel, simultaneously operating the engine throttle, until the desired load and speed are obtained. Adjust the outlet valve to pass sufficient water to keep the temperature at a reasonable figure of 60°C. When running very light loads (e.g., the lower worldwide mapping points [15 BMEP psi at 1500 rev/min crank] with the sluice gates fully closed), a further reduction in load may be obtained by opening the outlet valve and gradually closing the water inlet valve.

Hydraulic Water Brakes

There are two types of hydraulic water brakes: (1) the constant fill type, and (2) the variable fill type.

Constant Fill Hydraulic Water Brakes

This type uses thin sluice plates inserted between the rotor and the stator, across the mouth of the pockets, to interrupt and affect the development of the toroidal (or whirlpool effect) flow patterns within the pockets. These sluice plates can be inserted to infinitely varying degrees to provide variable control of the loading of the engine crankshaft.

Variable Fill Hydraulic Water Brakes

As the name suggests, this method relies on controlling the amount of water available within the dynamometer casing, thus affecting the water supply available to the rotor/stator assembly. This in turn will have an effect on the developed resistance force. The use of water outlet valves in varying the water flow through the dynamometer casing replaces the sluice plate control found in the constant fill machines.

A question arises here: Why is it that the flat-out maximum speed engine does not generate a resistant force in the dynamometer that is large enough to stall the engine? In some cases, it may indeed be possible to stall the engine with a high enough rating of dynamometer in good condition; however, this is not necessarily a desirable feature and is dangerous and damaging to both the engine and the dynamometer. When dynamometers reach this sort of range, damaging cavitation and erosion of the internal components can occur, particularly on the rotor and stator pocket faces. To prevent this, a means of controlling the resistance of the dynamometer to the torque generated by the engine is necessary. This controlling mechanism is the primary difference between the two types of water brake dynamometers.

Water brakes, as they are known, generally are being phased out in favor of electric dynamometers for their greater sensitivity of control. However, some remain in use, and they are helpful in developing the technician's understanding of engine loading for test purposes.

Electric Dynamometers

The two main types of electric dynamometers are the eddy current dynamometer and the AC or DC transient dynamometer (generator dynamometer). This section on electric dynamometers will focus on the eddy current type of dynamometer (Figures 1.15 through 1.17), with some comment on the use and limitation of the other electric dynamometers.

How the Eddy Current Dynamometer Works

With the eddy current dynamometer, the engine turns on the driveshaft, which is mounted on the rotor within the dynamometer casing. The outer edges of the rotor disc run between electromagnetic poles of the stator. Varying the excitations of these magnets, thereby altering their effect on the spinning rotor disc, will develop a resistant force or drag to counter the torque of the engine. Electromagnetic devices such as this are infinitely variable and have the added advantage of almost instantaneous implementation, thus giving greater control for the test-bed controller. Water cooling is achieved by passing

Figure 1.15 An eddy current unit. (Courtesy of Froude Hoffman)

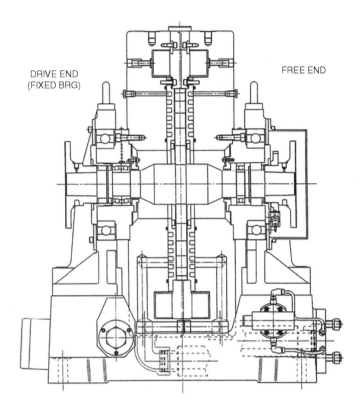

Figure 1.16 Drawing of an eddy current dynamometer. (Courtesy of Froude Hoffman)

Figure 1.17 An eddy current dynamometer installation. (Courtesy of Froude Hoffman)

raw water into the cavities in the stator near the point where the rotor and the stator are closest (i.e., where the magnets act upon the rotor plate). Eddy current dynamometers are the most popular type used within the test cell environment. These vary in size and application, depending on the power/torque output of the engine being run. Special care must be given to the elimination of vibration when using electric dynamometers because the sensitivity of the control will be affected.

14 An Introduction to Engine Testing and Development

The AC or DC Transient Dynamometer

The AC or DC transient dynamometer (Figures 1.18 and 1.19) consists of a variable speed generator or alternator, the electrical output of which is delivered outside the test cell to a controllable load bank. In certain cases, particularly where large power output and continuous operation are concerned, the output may be delivered into the mains supply. The generator cooling air is drawn from the test cell and returned to it, contributing to the total heat release in the cell. No water cooling is likely to be involved. This type also can be used for motoring the engine and is used primarily for transient testing. Here, the aim is to evaluate fuel consumption variances (among other values),

Figure 1.18 A 130-kW AC dynamometer. (Courtesy of Froude Hoffman)

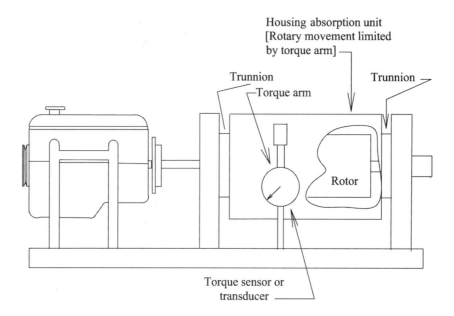

Figure 1.19 A transient dynamometer.

during split-second operations such as gear changing and overrun situations, which also affect fuel use and so forth, via the engine management systems.

Dynamometers must be kept within their specified calibration schedules because the accuracy of the dynamometers form the basis upon which all other data are generated and referenced. Dynamometer calibration information should be displayed where the test technician and the engineer can see it; this gives confidence to all, including the client's representatives, that best practices are being followed. Calibration usually is carried out by support technicians and involves the use of known mass weights placed on a specific-length calibration arm.

Mechanism of the Dynamometer

Figure 1.20 illustrates the operating principle of the dynamometer.

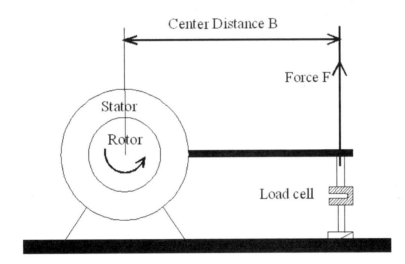

Figure 1.20 Torque measurement.

The rotor is coupled electromagnetically, hydraulically, or by mechanical friction to a stator that is supported in low-friction bearings. The stator is balanced with the rotor stationary (static calibration). The torque exerted on the stator with the rotor turning is measured by balancing the stator with weights, springs, or pneumatic means.

If the torque generated by the engine is T, then

$$T = Fb$$

The power P delivered by the engine and absorbed by the dynamometer is the product of torque and angular speed as

$$P = 2\pi NT$$

where N is the crankshaft rotational speed. In SI units,

$$P(kW) = 2\pi N(\text{rev/sec}) \times T(Nm) \times 10^3$$

or in U.S. units,

$$P(\text{hp}) = \frac{N(\text{rev/min}) \times T(\text{lbf} \cdot \text{ft})}{5252}$$

Note that torque is a measure of the ability of an engine to do work; power is the rate at which the work is done. The value of engine power as described here is called brake power, Pb. This power is the usable power delivered by the engine to the imposed load (i.e., brake power).

In essence, the dynamometer sets up a resistance to the rotating force (or torque) of the engine crankshaft. This resistance, in effect, is applying a load on the engine, making it work harder to maintain its rotational speed.

Let us now examine the dynamometer in greater depth and undertake a study of the performance or absorption curves of the differing types of dynamometer.

Figure 1.21 shows the four quadrants in which a dynamometer may be called to operate. The majority of engine testing takes place in the first quadrant with the engine running counterclockwise when viewed from the flywheel end. All types of dynamometers are naturally designed to run in the first or second quadrant.

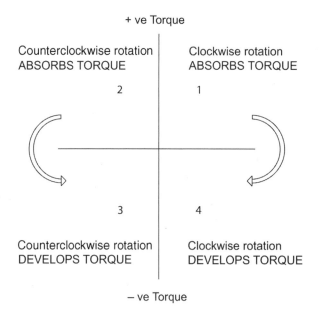

Figure 1.21 Dynamometer operating quadrants.

Table 1.2 shows a summary of the operating quadrants. Hydraulic dynamometers are designed in the main to run in only one direction, but they can be run in reverse without damage. As the name suggests, this method relies on controlling the amount of water available within the dynamometer casing, thus affecting the water supply available to the rotor/stator assembly. This in turn will have an effect on the developed resistance force. The use of water outlet valves to vary the water flow through the dynamometer casing replaces the sluice plate control found in the constant fill machines.

The Test Facility and Methods of Measuring Engine Power 17

TABLE 1.2
SUMMARY OF OPERATING QUADRANTS

Type of Machine	Operating Quadrant(s)
Hydraulic sluice plate	1 or 2
Variable fill hydraulic	1 or 2
Hydrostatic	1, 2, 3, 4
DC electrical	1, 2, 3, 4
AC electrical	1, 2, 3, 4
Eddy current	1 and 2
Friction brake	1 and 2

Dynamometer Characteristics

Figures 1.22, 1.23, and 1.24 show the characteristics of hydraulic, eddy current, and AC or DC dynamometers, respectively.

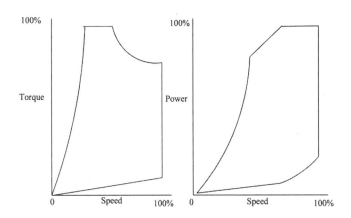

Figure 1.22 Hydraulic dynamometer absorption curve.

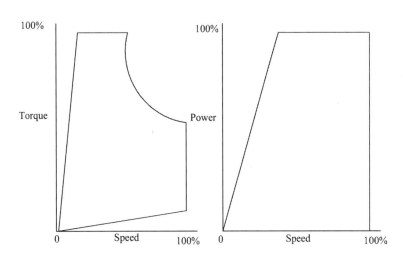

Figure 1.23 Eddy current dynamometer. (Courtesy of Froude Hoffman)

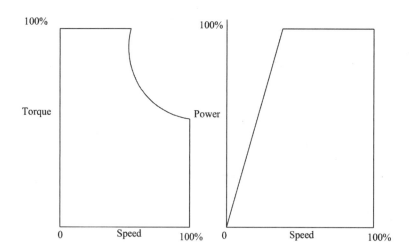

Figure 1.24 AC/DC transient dynamometer absorption curve.

The characteristics of the hydraulic dynamometer are as follows:

(a) Full maximum water. Torque increases with the square of the speed. No torque at rest.

(b) Performance limited by maximum permitted shaft torque.

(c) Performance limited by maximum permitted power, which is a function of cooling water throughput and its maximum permitted temperature rise.

(d) Maximum permitted speed.

(e) Minimum torque corresponding to minimum permitted water flow.

Figure 1.25 shows a general arrangement of the classic DPX hydraulic water brake.

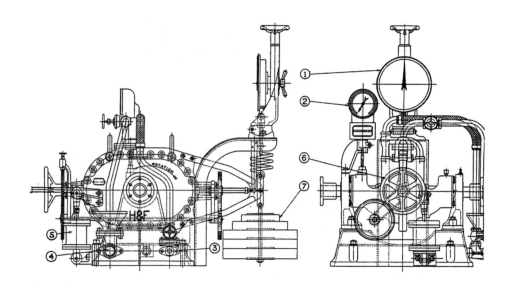

Figure 1.25 DPX hydraulic water brake arrangement. (Courtesy of Froude Hoffman)

The following are characteristics of the eddy current dynamometer:

(a) Low-speed torque corresponding to maximum permitted excitation.

(b) Performance limited by maximum permitted shaft torque.

(c) Performance limited by maximum permitted power, which is a function of cooling water throughput and maximum permitted temperature rise.

(d) Maximum permitted speed.

(e) Minimum torque corresponding to residual magnetization, windage, and friction.

The following are characteristics of the AC or DC transient dynamometer:

(a) Constant torque corresponding to maximum current and excitation.

(b) Performance limited by maximum permitted power output of the machine.

(c) Maximum permitted speed.

These performance curves are known as absorption or dynamometer envelopes.

The engineer will use this data when selecting a dynamometer for specific engine performance. For performance development, when best possible accuracy is required, then it is desirable to choose the smallest machine that will cope with the largest engine to be tested. For durability and validation work, it is best to select a machine that is rated approximately 50% above the maximum rated power of the engine in question.

Selection of Prop Shafts

In general, dynamometer shafts are not designed to withstand heavy bending moments, which can be caused by the use of heavy couplings or by misalignment between the dynamometer and the engine. For this reason, flexible couplings should be used of the lightest possible construction and in perfect dynamic balance. Cardan shaft or hook joint shafts frequently are a favored choice of automotive test engineers, preferably with a universal joint at each end of the shaft designed to run out by 2 to 4° to stop brinelling of the roller bearings in the joint cruciforms (Figure 1.26).

In the absence of a cardan shaft, alignment between the engine and dynamometer must be carried out with a high degree of accuracy. However, remember that the engine mounts will tend to move and settle after the initial installation, and the alignment will need to be checked regularly over the first few days of testing. Poor attention to prop shaft selection and installation are potentially dangerous to test cell personnel due to the high speeds and high inertia forces.

Suitability of a coupling should include the following considerations:

1. The ability to handle the peak instantaneous cyclical torque due to the engine under consideration.

2. The effect of the torsional stiffness of the coupling proposed, usually as a result of natural torsional frequencies of the installation in which the coupling is fitted.

3. The effect of the considerations in items 1 and 2 on the amount of heat generated in the coupling. This is of particular importance when rubber in shear or in direct tension or compression type couplings are considered.

Figure 1.26 Example of a hook joint cardan shaft. (Courtesy of University of Sussex)

4. The effect of the coupling so far selected on the whirling characteristics of the whole coupling shaft.

5. Possible detrimental effects due to the inability of the coupling to deal with reasonable run out. The effect of the coupling so far selected on the whirling characteristics of the whole coupling misalignment or, in certain cases, the effect of end loading produced as a direct result of torsional twist on the bearings and so forth of the engine and dynamometer.

Practical experience shows that successful operation of test bed couplings requires the cooperation of all personnel involved. The selection of appropriate couplings for a given installation is important; however, for example, the machinist must take care that the correct depth and diameter are provided for any spigot fits. Installation fitters must conform to the degree of alignment required between the engine and the dynamometer shaft. It must be ensured that the correct bolts are fitted and that they are tightened appropriately to the required torque. In addition, there are always critical frequencies that should be avoided: the test personnel must not run at barred speeds for prolonged periods.

It is wise to always fit a robust safety containment guard should there be an unforeseen catastrophic coupling failure, for the kinetic energy in a swirling broken propeller shaft is immense.

Chapter 2

In-Cell Services

The efficiency of a test cell is dependent on the provision of precision temperature control services, such as raw water and air supplies. Engines often need to be tested at extremes of temperature, perhaps as low as −35°C and as high as +46°C. The test cell sometimes can be a very uncomfortable place in which to work!

This chapter deals with raw water services and air services for both the test cell and the site.

Raw Water Services (Cooling)

For engines undergoing testing in an engine laboratory, it is usual to provide a common cooling supply to all the test cells in the laboratory area via a site raw water distribution system. An example of a test-cell cooling water circuit is shown in Figure 2.1 and is described more fully later.

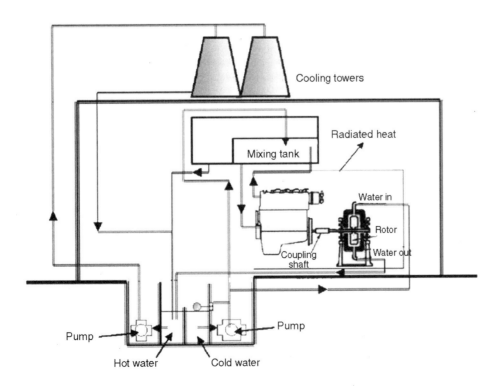

Figure 2.1 A test cell cooling water schematic.

Function of the Raw Water Cooling Circuit

The hot and cold wells provide a volume of water required for the running of numerous engine and dynamometer rigs under test at the same time. (Dynamometers are cooled by the water or air.) The wells provide the means for an infinitely variable and controllable engine heat range by controlling the flow and temperature of the raw water through the cell manifolds. This system is designed to provide instant engine cooling for repeated cold-start condition testing by rapidly dissipating the heat of the engine coolant at the heat exchangers. (This "crash" cooling is achieved by integrating a thermal shock rig or crash-cooling rig into the water supply at the cell.) The system reduces the requirement for manufacturer-supplied radiators, as heat exchangers generally replace the radiator in the engine coolant system.

Current test-cell applications set out to replicate the in-vehicle installation as much as possible, and in these cases, the vehicle radiators are utilized. Where ultra-cold running conditions are required (–25 to –35°C), a system of cold-fluid heat exchangers controls the coolant, oil, and fuel temperatures. An environmental chilling system dissipates the radiated heat within the cell, and a temperature-controlled air inlet system is incorporated. This type of system enables mapping of transient conditions to be undertaken in sub-zero conditions. Figure 2.2 shows an example.

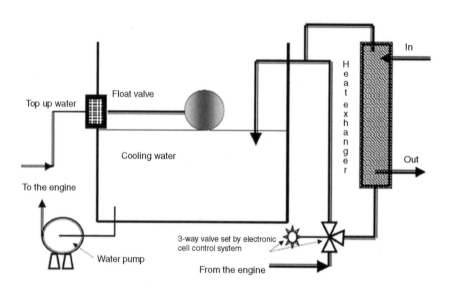

Figure 2.2 Example of a cooling water service module with a heat exchanger.

Site Water Services

Raw cooling water is stored below ground level in a split sump system. One-half of the sump (i.e., the cold well) stores cold water for supply into the test cells. The other half (i.e., the hot well) receives the hot used water back from the test cells and holds it ready for distribution to the cooling towers (Figure 2.1). The split sump has several features in addition to the two water storage areas:

- A mains water top-up system using a float valve allows fresh mains water to top up the split-sump cold well as is required.

- A dividing wall between the hot and cold wells is slightly lower than the normal water level to allow a bleed across the two wells. The dividing wall also has a small opening at the base to allow draining.

- A bleed-to-drain system is installed to allow a nominal amount of cooling water to leak away to the drain, usually about 1% of system capacity per day.

Cooled water is pumped to the test cell area via a delivery pump and suitable pipe-work (pressure and flow regulated). At the test cell area, it can be branched off into individual test cells and, once there, via a water manifold to the different parts of the engine and dynamometer rig assembly.

Individual cells can be isolated for maintenance and new installations by fitting shutoff or isolation valves at strategic points in the supply side of the cooling system, as shown.

With reference to Figure 2.1, used water flows from the engine and dynamometer to an opposite manifold in the cell and then into a common return pipe or underground channel, where it is directed back to the hot well in the split sump. Once in the hot well, it is pumped to the cooling tower by the transfer pump. Then it enters the tower at the top and cools as it falls through the tower body. The tower cooling temperature is controlled by a thermostat, which operates and controls the speed of huge cooling fans drawing air across the tower body assembly. Cooled water then is returned to the cold well area of the split sump, ready for recycling.

Water is lost through evaporation due to high temperatures in the system and through deliberate leaking. This deliberate leaking is done to ensure that the cooling water is recycled with fresh water over a period of time to prevent excess buildup of impurities and water quality deterioration. The bleed-off reduces the requirement to drain and replenish the system completely on a regular basis.

The use of water cooling towers is prone to the breeding of bacteria that are dangerous to human beings. For example, Legionnaires' disease is a deadly condition that is introduced by inhaling airborne bacteria given off by the water in the cooling tower system. It is absolutely imperative that expert help be obtained from water treatment specialists by any company operating test-cell cooling towers.

Likewise, it is important to ensure that the system is designed to adequately cool the envisaged range of engines to be tested into the foreseeable future. In the case of dynamometers, the circulating water should be treated to ensure the minimum buildup of hard scale calcite deposits within the dynamometer. (This condition is accelerated if the dynamometer cooling water temperature is high.) The recommended dynamometer outlet temperatures are 60°C for eddy current machines and 65°C for hydraulic units. (A temperature of 70°C can cause localized boiling and cavitation in the casings of hydraulic dynamometers.)

Flow rate to and from the dynamometer is vital to protect and preserve the life span and serviceability of the dynamometer. Water delivery pressure and stability must be ensured when using hydraulic dynamometers to avoid the loss of control at the dynamometer. Water quality and cleanliness are required to prevent the ingress of dirt and unwanted materials into the dynamometer casing. In many cases, water may require softening to reduce or prevent scaling within the dynamometer casings.

In-Cell Water Services

The cooling raw water delivered to the test cell area by the delivery pumps is received in the individual cell into a Hale-Hamilton type of valve. The water then goes on to a manifold, usually approximately 50 mm in diameter. The manifold is used to distribute the water around the various parts of the engine/dynamometer assembly.

Cooling water is required at the following points:

- Hale-Hamilton valve
- Dynamometer
- Heat exchangers
- Saunders valves

Engine Coolant

It must be emphasized here that the engine coolant (with or without antifreeze mixed into the coolant) is separate from the cooling raw water. The engine coolant is stored in a host radiator/chamber that also may incorporate the heat exchanger, with separate chambers keeping the engine coolant away from the cooling raw water. The components of the engine cooling system and their relative positions and connections are critical and must conform to the requirements laid down.

The cooling raw water carries away the engine coolant heat to the hot well and cooling towers. Frequently, a customer requests that his or her engine be run on a particular coolant and antifreeze mixture for certain tests. This mixture is set up for the coolant system of the engine, and the cell raw water services are used to maintain running temperatures as required by the test procedures. Note that all antifreeze solutions are not the same. Therefore, care should be used to ensure that technicians select the correct antifreeze, per the instructions of the engineer and client. This is particularly important when ultra-low emission levels are being measured, for the heat transfer rate through the mixture can affect the combustion process. In addition, the onset of nucleate boiling and effective heat transfer can be adversely affected by incorrect coolant mixtures, making repeat tests at some time in the future practically impossible.

The Hale-Hamilton coolant damper valve is incorporated into the raw water system on the inlet side before the water manifold. Its function is to smooth out the pump surges and to accumulate pressurized water in a pressure reservoir. This ensures that there is always a stable source of cooling water in a pressurized state for feeding to the dynamometer via the manifold in the test cell.

Heat Exchangers

The main function of heat exchangers is, as the name suggests, to exchange heat. It is important to remember that heat can be exchanged in both directions using the heat exchanger system.

The new generation of plate heat exchangers has been designed as a low-cost alternative to the shell/tube units. High-pressure devices are made from grade 316 stainless steel heat transfer plates, two outer covers, and four connections, with copper vacuum brazed together to form an integral unit. Low-pressure plate-type heat exchangers are manufactured from aluminum and duralumin. These types of heat exchangers are suitable for heating, cooling, evaporating, or condensing any fluids compatible with the materials of construction.

Heat is exchanged inside the body of the heat exchanger, between the two separated chambers, with one chamber housing the cooling or heating agent (raw water supply), and the other housing the agent (fuel, oil, antifreeze mix) to be cooled or heated.

Passing the cooling water through the small tubes in the core of the heat exchanger allows it to draw the heat from whatever is in the outer chamber (i.e., that surrounding the tubes) (Figure 2.3). If hot engine oil is passed into the outer chamber of the exchanger, and cold water is passed through the tubes, then the heat is transferred to the cold water. Thus, cooling of the engine oil is possible. Equally, if hot water is passed through the tubes, and cold engine oil is in the outer chamber, then heat is transferred the other way and the engine oil is warmed. Figures 2.3 and 2.4 show examples of heat exchangers, and Figure 2.5 shows a diagram of the core tubes.

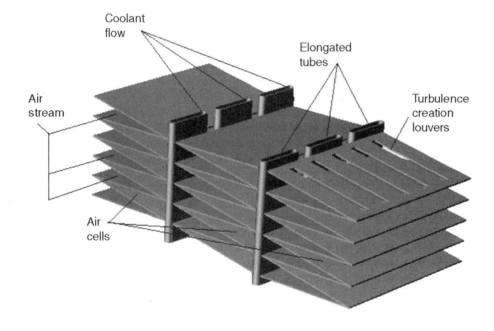

Figure 2.3 Elongated tube and flat fin heat exchanger.

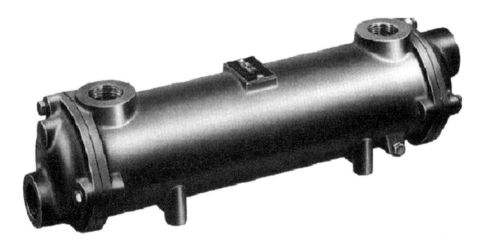

Figure 2.4 Typical heat exchanger. (Courtesy of E.J. Bowman [Birmingham] Ltd.)

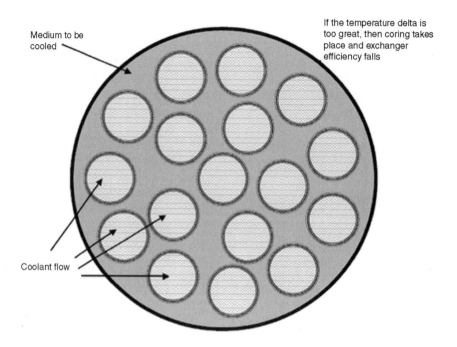

Figure 2.5 Heat exchanger core tubes.

Note that in the case of engine oil, if the temperature delta (Δ) is too high, then the oil will core, resulting in very fast high-temperature flow in the center of the tube, and the oil will cool and be near-stationary along the walls (Figure 2.6). Due to this coring effect, it is important to size the heat exchangers so that the temperature delta (Δ being the difference between the highest and the lowest recorded temperatures within the system) is below 25°C.

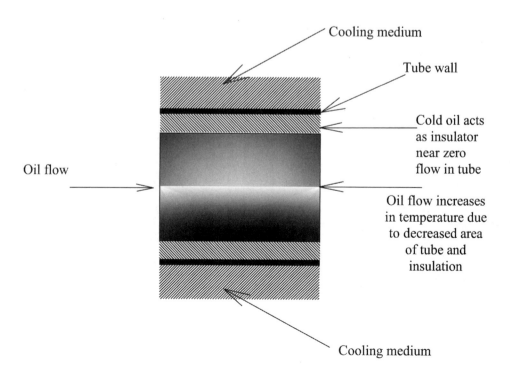

Figure 2.6 Tube coring.

In testing engines, an infinite variety of temperature conditions may be required, ranging from continuous hot or cold running to rapidly changing running temperatures. The heat exchangers can be used to control and stabilize the temperatures of engine oil, coolant/antifreeze, fuel, power steering fluids, and so forth (Figure 2.7).

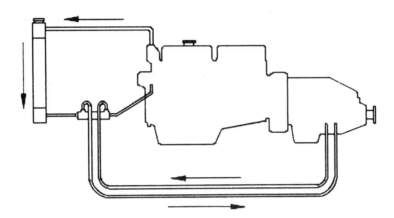

Figure 2.7 Position of a heat exchanger in a torque converter system.

Table 2.1 presents the oil and coolant flow required for the cooling of an automatic gearbox oil system.

TABLE 2.1
OIL AND COOLANT FLOW REQUIRED TO COOL AN AUTOMATIC GEARBOX OIL SYSTEM

Maximum Oil Flow (Liter/Min)	Maximum Water Flow (Liter/Min)	Suitable for Torque Converter Transmitting (kW)	Internal Oil Volume (Liter)	Internal Volume of Water (Liter)
30	200	45	0.26	0.31
60	200	60	0.49	0.44
60	200	75	0.74	0.57
60	200	90	0.97	0.71
60	200	105	1.3	0.91

Although generally reliable, heat exchangers can break down. Some obvious symptoms can be apparent in diagnosing breakdown, such as the following:

- The mixing of the two agents, for example, the emulsifying of oils and fluids as water enters oil systems.

- The pressurizing of normally low or unpressurized systems as the higher-pressure agent corrupts the lower-pressure agent through internal leakage in the heat exchanger.

- Fluid levels increasing in the engine systems as raw cooling water leaks into and tops up engine coolant or engine oil.

- Poorly running engines as cooling water enters the fuel system and corrupts the fuel quality. Likewise, levels in the closed engine systems decreasing as leakage occurs across the chambers in the exchanger body.

- External leakage of both chambers can occur at the sealed faces and connections.

Heat exchangers usually are made of a design that allows easy maintenance and repair of the individual chambers. Referring to Figure 2.8, removal of the end caps will allow access to the tubes, and the outer casing or body forms the outer chamber. The end caps and the chambers are sealed against each other on assembly. Sometimes a drain tap is fitted into the outer casing to enable drain down as required.

Figure 2.8 Heat exchanger sub-assembly. (Courtesy of E.J. Bowman [Birmingham] Ltd.)

Header Tank Heat Exchangers

These heat exchangers originally were designed for marine engines, but they also are widely used for various land-based duties, such as engine testing and development work, generator sets, fire pumps, and combined heat and power systems. They incorporate a header tank with a special de-aeration space and a pressurized filler cap. The removable tube stack is held in place with O rings and is free to expand and contract within the cast housing, thus minimizing thermal stresses. It can easily be removed for cleaning as required. When installing, the heat exchanger should be mounted with the header above the cylinder head level and with the engine water circuit arranged so that it is self-venting upon initial fill. A bypass type of thermostat can be used and arranged so that only the heat exchanger is bypassed when the engine is cold.

Note that thermostats of the type used on many in-vehicle automotive applications, which simply interrupt the cooling water flow when the engine is cold, are not recommended. For typical unattended running operations, automatic engine shutdown equipment should be provided.

Saunders valves are used to control the pressure of the cold water within the system and to dampen pressure surges that occur frequently within closed water systems. Three-way flow control valves actuated via air pressure and electronic control are used in conjunction with heat exchangers to control the temperature of coolant, oil, fuel, recirculated exhaust gases, or compressed induction air. Generally positioned at or near the outlet from the requisite heat exchanger, the valve restricts flow from the exchanger and therefore can retain heat within the heat exchanger until a required temperature is attained. The valve is controlled by the test-bed control equipment (a proportional integral derivative [PID] system, which will be discussed later in this chapter) and is either pneumatically or electrically activated and fails to a safe condition to prevent overheating. The control temperatures can be set on the test-bed equipment, and the Saunders valve responds to the control signals when required.

Note that the control temperatures for the test should be stated clearly in the test instructions.

Fuel, oil, or coolant is passed into the casing and imparts its heat to the raw cooling water flow in the tubes, which run from end to end. The raw cooling water runs through tubes absorbing heat from the fuel, oil, or coolant flowing around them.

Air Services

In dealing with air services to the engine test cell, there are two primary functions that the air provides for the engine: cooling and breathing. The control of one of them (cooling) may be critical to the effects that the other (breathing) will have on the engine. As a general rule, internal combustion engines are considered to be air pumps, in that they breathe in air for combustion purposes. Therefore, it follows that the quality of that air with regard to its temperature, pressure, humidity, and condition will affect the performance of the engine.

To illustrate the point, consider this scenario. Have you ever noticed that the engine of your own vehicle, be it car or motorcycle, seems to be more responsive during cold and damp weather conditions than on hot, dry days? Try driving late at night on a hot day, and compare the daytime performance with the night-time performance. Invariably, the engine will be more responsive under the colder conditions.

In the preceding illustration, any or even all of the four points mentioned (air temperature, pressure, humidity, or condition) could affect the performance of the engine. Because breathing and cooling are two separate functions, they are considered separately with references where necessary to the interaction between the two.

Combustion or Induction Air

Combustion or induction air describes the air that the engine will draw in, or induct, through the inlet tract into the inlet manifolds and into the cylinders for the combustion process. The source of this air usually is the ambient air that is present within the test cell itself (i.e., the same air that the technician is breathing—fumes, vapors, and all). The quality of the air is critical to the performance of the engine and will be reflected in the engine test results. Obviously, from the viewpoint of the customer who is commissioning the tests, undesirable air quality will produce undesirable test results and inevitably lead to wasted money, time, and effort in conducting the tests. Where in-cell air quality is of an extremely dubious or poor condition with regard to the four points mentioned (air temperature, pressure, humidity, or condition), or the test requires absolute guaranteed and consistent air quality, then the air for the combustion process may be drawn from outside the cell and via additional filtering equipment. In the main, air is drawn from the cell environment, and this is the focus of this section.

Air Temperature

The temperature of the air entering the inlet manifold has a direct influence on the complete integration of the air and fuel as a mixture, with regard to the evenness and atomization of the mixture (i.e., the dispersal and size of the fuel droplets throughout the air stream and on the mass of the air/fuel mixture). When the air temperature is too low, fuel does not mix as effectively with the air stream due to the higher density of the air; therefore, it tends to fall to the sides and floor of the inlet tract. This gives an uneven mixture and poor atomization, which may cause misfiring and so forth. In turn, this can lead to higher emissions of hydrocarbons and carbon monoxide. In the case of air temperatures that are too high, the air charge is expanded by its heat and, as a consequence, has a reduced density. This reduced density means that the fuel/air charge will have a lesser mass than preferred (lower total oxygen content).

Note that the power developed by the cylinder of an engine is proportional to the mass of the fuel/air mixture drawn into the cylinder. If the temperature of the fuel/air charge is too high, then the maximum power that the engine will achieve is reduced.

The optimum combustion or induction air temperature obviously lies someplace between the two extremes and may vary with the engine or management system fitted for the test purpose.

Vehicle induction tracts may have a temperature control device fitted into the air filter intake pipe and connected to the exhaust manifold heat shield, to transfer heat to the air filter temperature control device and to help maintain a consistent desired temperature range in the air entering the engine. The device usually is a flap operated by a temperature sensing mechanism that progressively lets hot circulated air (drawn from the area of the exhaust manifold) into the engine inlet tract to warm the induction air. The test engineer should ensure that, where applicable, the required air temperature figure is stated clearly in the test instructions before starting the testing; otherwise, the results might be questionable.

The engineer must ensure that the air inlet temperature remains constant throughout the test as required and should pay due regard to positioning of equipment that may affect the temperature control (e.g., spot coolers).

Measuring of the air temperature can be done by situating the sensing thermocouple either in the inlet tract or in the free space near the opening to the tract, where it will record the ambient air temperature in the area where air is drawn into the engine inlet tract.

Air Pressure

Combustion or induction air pressure has a direct bearing on the performance of the internal combustion engine. The mass of the fuel/air mixture charge within the inlet manifold ready to be drawn into the cylinder can be increased or decreased by a high or low air pressure, respectively. The power developed by the cylinder of an engine is proportional to the mass of the fuel/air mixture that is drawn into the cylinder. An example in appreciating the idea of air pressure is in the supercharging and turbocharging of engines. Both pressurize the air (among other things), forcing more air/fuel mixture (or air alone in the case of diesel engines) into the cylinders in the short time allowed by the inlet valve opening period, thereby increasing the mass of the charge and increasing the maximum attainable power.

If the test engine is inducting its combustion air from the ambient air in and around the test cell area, then it will be vulnerable to changes in the atmospheric conditions that affect air pressure. In many test cells, barometers are a common sight, and the former practice was that frequent readings of the ambient pressure would need to be recorded and then be taken into account when recording the test data. The current practice is to continuously monitor the air pressure within the cell, with automatic correction calculations done by the data acquisition equipment. Improperly sited spot coolers can have the effect of ramming the air into the air inlet tract if positioned too close to the opening to the tract. The ramming effect increases the turbulence of the air and thus its pressure. This may lead to inconsistent test results over a period of time, especially if the cooler is moved later during the test. Air turbulence at the induction area also can steal air from the induction process. Turbochargers and superchargers are introduced to increase the mass of the charge of air/fuel (or air only for diesels) entering the inlet manifold.

Note that in typical road vehicles, the air generally enters the inlet tract to the air filter under ram air conditions as the vehicles move along the road at speed, pushing through the surrounding air. Depending on the air speed, this effect normally is small (e.g., 0.1 bar at 60 mph).

Air Humidity

In general test cell operations, the humidity of the air admitted to the cell and entering the engine air inlet is not controlled (i.e., it is that prevailing in the general atmosphere). As legislated emission standards become more stringent, the control of humidity within the cell becomes a key issue because the degree of humidity affects the NOx emissions and the formation of particulate platelets in the combustion space. Both humidity and air induction temperature must be controlled in order to replicate tests at some time in the future. In the case of non-climate-controlled basic test cells, where air temperatures are concerned, there is quite a lot that the engineer can do to control the conditions within the test cell area. One example is ventilation.

Ventilation

The efficient flow of air through and around the test cell is important for the safety and reliability of the testing operation. Although airflow rate and extraction through the cell usually are controlled automatically, the flow around the assemblies and equipment within the test cell is affected by the care and thought applied by the technician when setting up the test. It is important that cells are kept tidy and clear of unnecessary equipment, particularly near the ventilating fans and so forth.

Electrical Equipment

With so much sensitive equipment located in and around the test cell, individual appliance positioning is vital, particularly where numerous items of electrical equipment are located together, each generating its own heat outputs. Most electrical appliances usually have a ventilating panel to aid cooling of the components inside. Some thought should be given as to how best to position these appliances to help their cooling, thereby reducing the potential for hazards and faulty equipment.

The fact is that electrical appliances cause most in-cell fires due to faulty cabling, incorrect installation, poor ventilation, or simple misuse and abuse!

With regard to basic test cells, the problem for the engineer is to understand and remove, as much as possible, the potential causes of increases in humidity. The technician also should consider humidity. The local buildup of humidity to excessive levels would have detrimental short- and long-term effects on electrical-equipment-measuring instrumentation being used during the tests. Failure to appreciate this fact can lead to cases of expensive equipment failures, as well as time wasted in analyzing test results that owe more to faulty instrumentation than to the characteristics of the engine. When testing prototype engines and components, total confidence in the measurement instrumentation is of the highest importance. Awareness of the effects of excessive temperatures in the cell, and the further complications this may cause, will help to minimize potential problems.

Air Condition (Quality)

Thought also should be given to the effect on air quality of fuel and exhaust leakages within the test cell area, which are not only detrimental to the test results but also are potentially dangerous and even lethal to the technician.

As would be expected in an industry that tests engines to extremely high specifications and uses expensive and sensitive equipment (not to mention expensive engines), the need for vigilance in all aspects of testing is of the utmost importance (e.g., good fuel, good conditions, good practices, good technicians).

Air Cooling and Ventilation

Air as a cooling medium is not as effective as a liquid cooling system, particularly where the cooling of hot surfaces is concerned, as those surfaces are in engines and dynamometers running flat-out under load. The principal reasons for this are that air generally has a lower density than liquid and is considered to be invisible to radiant heat; in effect, this means that the air has poor heat transfer qualities under normal circumstances. However, with high-velocity air, as in spot cooler fans, the heat transfer rises considerably, as one would expect. But beware of the problems that spot cooler fans can cause, as discussed in the heat exchanger section, particularly with regard to position and air turbulence near the inlet tract intake.

Consider in the following list the heat output that each will contribute within a test cell, where the engine may run at full speed/full load for lengthy periods:

- Engine
- Exhaust
- Dynamometer—AC or DC motors
- Shaft and couplings
- Electric lighting
- Spot cooler motor
- Ventilator fans and motors

All of these contribute to making the test cell a very hot place in which to operate at times. Good, effective ventilation and cooling are vital for safety reasons as well as for test result reliability. Likewise, engineers and operators require a convivial work place, as shown in Figure 2.9.

Figures 2.10 and 2.11 illustrate the various types of testing from anechoic to large engine test work.

The main components of the process enclosure (Figure 2.12) of the fuel temperature conditioning unit are as follows:

Figure 2.9 Control area of an engine test facility. (Courtesy of Froude Hoffman)

Figure 2.10 Anechoic test cell. (Courtesy of Froude Hoffman)

- Circulating pump
- Selector valves
- Header tank
- Immersion heater
- Flat plate heat exchanger fed by chilled water circuit
- Temperature and level sensors

The control enclosure marshals the mains switching and input/output (I/O) data transfer associated with the control of the process enclosure. The process enclosure requires one set of feed and return lines for the process water circuit and one for the chilled water circuit. This minimizes the pipe-work in the test cell. The temperature in the header tank is maintained or adjusted by directing the intake closed loop water.

Docking Rigs

Docking rigs are used in end-of-line production test applications where rapid turnaround of engines is a prerequisite (Figures 2.13 and 2.14). Typical times for installation run-up under power, engine condition diagnosis, and removal usually are less than 10 minutes. Hot fluids frequently are used to speed the operation.

Figure 2.11 A large engine on test. (Courtesy of Froude Hoffman)

Figure 2.12 Fuel conditioning equipment.

In-Cell Services 35

Figure 2.13 Automatic docking rig. (Courtesy of Froude Hoffman)

Figure 2.14 An engine on an automatic docking rig. (Courtesy of Froude Hoffman)

Some Engine Testing Pointers

Pre-Start, Operating the Dynamometer

Water Dynamometer

Prior to starting the engine, open the inlet valve fully and the outlet valve slightly. It is almost always advisable to start with a light load, and this may be accomplished by screwing the sluice gates into the dynamometer as far as they will go. The engine now is started. To regulate the load, open the sluice gates by means of the handwheel, simultaneously operating the engine throttle, until the desired load and speed are obtained. Adjust the outlet valve to pass sufficient water to keep the temperature at a reasonable figure of 60°C. When running light loads, such as the lower worldwide mapping points (very low speed and very low levels of developed torque such as 1500 rev/min at 15 BMEP psi), with the sluice gates fully closed, a further reduction in load may be obtained by opening the outlet valve and gradually closing the water inlet valve.

Electric Eddy Current and AC/DC Dynamometer

Ensure that if running in constant torque mode, you start the engine with zero or near-zero torque position selected on your test-cell control system.

Predictive Analysis

When testing engines, the ability to predict when a component will fail due to fatigue or wear-out is a prerequisite. This prediction can be brought into play by accelerated rig testing of key components; utilizing intelligent design validation test methods, with much attention paid to pre- and post-test strips; examination and measurement of critical components; and the taking of regular samples of oil for further analysis and exhaust gas particulate speciation. These all are valuable tools used in predictive analysis studies.

A typical example is a 1000-hour validation test:

- On the post-test strip and measurement, it was noted that one cylinder bore had indications of wear and scuffing.

- It also was noted that the particular cylinder top compression ring was worn.

- Analysis of the oil samples taken at 50-hour intervals clearly showed a high level of chromium and iron in the first 50-hour sample; thereafter, no significant levels were noted.

- From this, the engineer deduced that the cause was ingress of foreign matter in the top piston ring groove that was dislodged at some time within the first 50 hours of running.

- However, if there had been chromium and iron in all 20 oil samples, then it would be possible to predict the projected life of the bore. Indeed, one would be able to alert the chief engineer to a possible coolant circulation problem leading to cylinder bore distortion.

A Transient Test

The salient features of a typical transient test bed application for a "World Rally Championship Engine" are as follows:

- Maximum speed 10,000 rev/min
- Maximum torque 1000 Nm
- Speed gradient 10,000 rev/min/second
- Fuel measurement 180 liters/hour
- Cooling capacity 600 kW
- Cycle simulation 10 Hz
- Standard data acquisition (96 channels, 50 Hz)
- High-speed data acquisition (16 channels, 50 Hz, 6-speed channels, 10 GHz)
- Cylinder pressure indication (4 channels, 0.05° crankshaft [crank angle])

Figure 2.15 shows dynamometer control readouts.

Figure 2.15 Dynamometer control readouts. *(Courtesy of University of Sussex)*

The Key to Control Systems

Proportional integral derivative (PID) control is the most common algorithm used in industry. As the name suggests, the PID algorithm consists of the three basic coefficients of proportional, integral, and derivative, which are varied to obtain the optimal response. PID is used as a means of continuous automatic monitoring and adjustment. The basic idea behind a PID controller is to read a sensor, then compute the desired actuator output by calculating proportional, integral, and derivative responses, and then sum those three components to compute the output.

Control loop makes it possible for a control system to remain at or near a specified setpoint. Proportional, integral, and differential gain (PIDG) is the term applied to control loop gains and stability settings. In a typical control system, the process variable is the system parameter that needs to be controlled, such as temperature, pressure, or flow rate. A sensor is used to measure the process variable and to provide feedback to the control system. The set point is the desired or command value for the process variable, such as temperature, speed, or torque.

At any given moment, the difference between the process variable and the set point is used by the control system algorithm (compensator) to determine the desired actuator to drive the system. For example, if the desired coolant temperature set point was 98°C and the actual coolant temperature was 1200°C, then the actuator output specified by the control algorithm would drive a tap to allow more cooling water to pass through a heat exchanger, resulting in a decrease in the temperature process variable. This is called a closed-loop control system because the process of reading sensors to provide constant feedback and calculating the desired actuator output is repeated continuously and at a fixed loop (Figure 2.16). In some instances, the actuator is not the only signal that has an effect on the system. For example, in a temperature-controlled test cell, there might be a source of cold or hot air that sometimes blows into the cell and disturbs the temperature. This is referred to as a disturbance. One normally tries to design the control system to minimize the effect of disturbances on the process variables.

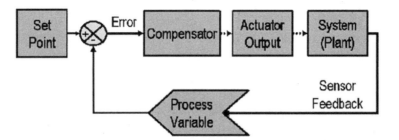

Figure 2.16 Block diagram of a typical closed-loop system.

The control design process begins by defining the performance requirements. Control system performance often is measured by applying a step function as the set point command variable and then measuring the response of the process variable. Commonly, the response is quantified by measuring defined waveform characteristics. Rise time is the amount of time the system takes to go from 10 to 90% of the steady-state or final value. Percentage overshoot is the amount that the process variable overshoots the final value, expressed as a percentage of the final value. Setting time is the time required for the process variable to settle within a certain percentage (in the automotive research and development field, this is commonly 5%) of the final value. Steady-state error is the final difference between the process variable and a set point. The exact definitions of these quantities will vary within the industry and academia.

After using one or all of these quantities to define the performance requirements for a control system, it is useful to define the worst-case conditions that the control system will be expected to handle. Frequently, a disturbance in the system affects the process variable or the measurement of the process variable. It is important to design a control system that performs satisfactorily during worst-case conditions. A measure of how well the control system is able to overcome the effects of disturbances is referred to as the disturbance rejection of the control system. In some cases, the response of the system to a given control output may change over time or in relation to some variable.

A nonlinear system is a system in which the control parameters that produce a desired response at one operating point might not produce a satisfactory response at another operating point. For instance, a chamber partially filled with fluid will exhibit a much faster response to heater output when nearly empty than when it is full of fluid. A measure of how well the control system will tolerate disturbances and nonlinearity is referred to as the robustness of the control system.

Some systems exhibit an undesirable behavior called dead time. Dead time is a delay between the time when a process variable changes and the time when that change can be observed. For example, if a temperature sensor is placed far from a cold water fluid inlet valve, it will not measure a change in temperature immediately if the valve is opened or closed. Dead time also can be caused by a system or output actuator that is slow to respond to the control command (e.g., a valve that is slow to open or close). A common source of dead time is fluid flow through pipes when thermocouples have not been sighted intelligently.

Loop cycle also is an important parameter of a closed-loop system. The interval of time between calls to control the algorithm is the loop cycle time. Systems that change quickly or that have complex behavior require faster control loop rates.

Once the performance requirements have been specified, it is time to examine the system and to select on appropriate control scheme. In most applications, a PID controller will provide the required results.

Proportional Response

The proportional component depends on only the difference between the set point and the process variable. This difference is referred to as the error term. The proportional gain (K_c) determines the ratio of output response to the error signal. For instance, if the error term has a magnitude of 10, a proportional gain of 5 would produce a proportional response of 50. In general, increasing the proportional gain will increase the speed of the control system response. However, if the proportional gain is too large, the process variable will begin to oscillate. If K_c is increased further, the oscillations will become larger. Then the system will become unstable and may even oscillate out of control.

Integral Response

The integral component sums the error term over time. The result is that even a small error term will cause the integral component to increase slowly. The integral response will increase continually over time unless the error is zero; therefore, the effect is to drive the steady-state error to zero. Steady-state error is the final difference between the process variable and the set point. A phenomenon called integral windup results when integral action saturates a controller without the controller driving the error toward zero.

Derivative Response

The derivative component causes the output to decrease if the process variable is increasing rapidly. The derivative response is proportional to the rate of change of the process variable. Increasing the derivative time (T_d) parameters will cause the control system to react more strongly to changes in the error term and will increase the speed of the overall control system response. Most practical control systems use very small derivative time (T_d) because the derivative response is highly sensitive to noise in the process variable signal. If the sensor feedback signal is noisy or if the control loop rate is too slow, the derivative response can make the control system unstable.

Tuning

The process of setting the optimum gains for PID to obtain an ideal response from a control system is called tuning. The gains of a PID controller can be obtained by trial and error. Once an engineer understands the significance of each gain parameter, this method becomes relatively easy. In this trial-and-error method, the I and D terms are first set at zero, and the proportional gain is increased until the output of the loop oscillates. As the proportional gain increases, the system becomes faster. However, care must be taken not to make the system unstable. Once P has been set to obtain a desired fast response, the integral term is increased to stop the oscillations. The integral term reduces the steady-state error but increases overshoot. Some amount of overshoot is always necessary for a fast system so that it can respond to changes immediately. The integral term is tweaked to achieve a minimal steady-state error, and the derivative term is increased until the loop is acceptably quick to its set point. Increasing the derivative term decreases overshoot and yields higher gain with stability but would cause the system to be highly sensitive to noise. Engineers frequently need to trade one characteristic of a control system for another to better meet requirements.

Chapter 3

Instrumentation: Temperature, Pressure, Flow, and Calibration

This chapter is concerned with the importance of instrumentation with regard to temperature, pressure, flow, and calibration. In all engine test work, instrumentation is paramount. Unless we measure and control various parameters, it will be impossible to replicate tests at some time in the future. There is no such thing as a finite measurement, be it temperature, pressure, or flow. An understanding of the uncertainty of measurement will ensure that results are as accurate as possible.

Temperature: The Principle and Application of Thermocouples

In 1821, Thomas Johann Seebeck, a German-Estonian physicist, discovered that if heat is applied to a junction of two metals that form part of an electric circuit, a current flows in the circuit. However, in 1834, Jean Peltier, a French physicist, demonstrated that if a current is passed through a junction of two dissimilar metals, heat is either absorbed or liberated, according to the direction of the current. The two discoveries are described as the Peltier-Seebeck effect.

The Principle of Thermocouple Operation

If two different metal strips or wires are joined together at both ends and those two junctions are at different temperatures, then electric current will exist (Figure 3.1).

The level of voltage produced is determined by the types of metal in combination and the differential in temperatures at each junction. Therefore, a thermocouple can measure only the temperature difference between those two junctions.

- A thermocouple is any pair of dissimilar electrically conducting materials coupled at an interface.

- A thermo-element, or leg, is an electrically conducting material used to form a thermocouple.

- An electrical interface between dissimilar electric conductors is a thermoelectric junction.

- A free end of a thermo-element is a terminus.

- Couplings between identical thermo-elements are joins.

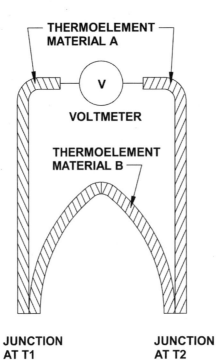

Figure 3.1 Thermocouple circuit.

The Law of Interior Temperatures

The thermocouple is unaffected by hot spots along the thermo-element, and the reading is only a function of T_1 and T_0 (Figure 3.2).

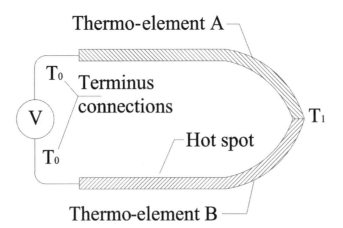

Figure 3.2 Hot spot.

Should there be a suspect reading or no reading at all shown on the data acquisition (DAQ) system, then the polarity of the feed wires should be checked (Figure 3.3).

A mineral-insulated metal-sheathed (MIMS) thermocouple is selected for its very high temperature capability and overall protection from the local environment (Figure 3.4).

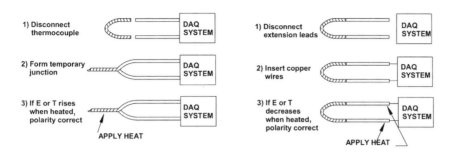

Figure 3.3 Checking polarity.

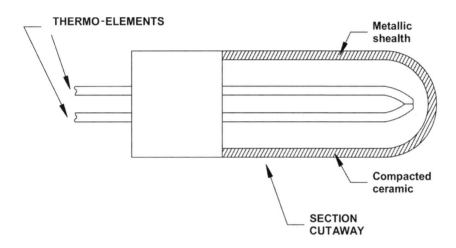

Figure 3.4 MIMS thermocouple.

One of the junctions is called the cold junction. This is the reference point that is kept at a known temperature and is used to determine the temperature at the other end (the hot junction), based on the voltage generated. This thermoelectric voltage is evaluated in context with known constant values associated with the types of metal in combination and their characteristics. The opposite end of the cold junction, or the reference junction, is the hot junction or the measuring junction. This junction is placed in the material or gas that is being measured (e.g., exhaust, coolant, oil gallery). Note that correct locating of this end within the area being measured is crucial for accuracy reasons. The whole device, including both junctions and all that lies between them, is considered to be the thermocouple. Because thermocouples are made by joining two different metals under a temperature differential at their junctions, care should be given to extending the thermocouple wires to reroute them or to reach other measuring and display equipment. Using another type of metal (e.g., copper extension cables connected to a non-copper thermocouple) will form a secondary thermocouple; therefore, corruption of the measured temperatures will be inevitable. Thus, it is important to use only correct compensating cables that are compatible with the thermocouple being used. This also implies that correct cable length must be used; therefore, shortening of the cable is not an approved technique, particularly because it will negate the prior calibration of the thermocouple.

Cold Junction Compensation

A primary issue with thermocouple use is maintaining the temperature as a constant at the cold junction or reference point, where it connects to the copper wiring in the display or data acquisition equipment. Such equipment usually has a compensating device that will sense the temperature at the cold junction connection to the equipment and send a corrective signal to compensate for this.

Hot Junction Protection

The measuring tip of the thermocouple usually is protected from the effects of flow, pressure, and corrosive sources at the point being measured by encasing it in a sheath. There are three types of sheathing arrangements, and they are selected depending on what and how measurements are being taken.

- **Exposed (measuring) junction**—An exposed junction is used for measuring flow-in or static non-corrosive gas temperatures where greater sensitivity and responsiveness are important.

- **Insulated junction**—An insulated junction is more suited for use with corrosive media with a slower response time because the junction is not in direct contact. It also should be used when several thermocouples are connected to one measuring/display device to reduce the risk of cross interference through spurious signals between the thermocouples.

- **Earthed junction**—An earthed junction also is used for measuring in corrosive media but has higher responsiveness than the insulated type. At the same time, an earthed junction is more protected than the exposed type.

Standardized Thermocouples and Categories

There are eight standardized thermocouples, which fall into three general categories:

- Rare metal thermocouples (types B, R, and S)
- Nickel-based thermocouples (types K and N)
- Constantan negative thermocouples (types E, J, and T)

Table 3.1 gives specifications for thermocouples.

Pressure: A Review of Pressure Measuring Devices

What Is Pressure?

For a thermodynamic system consisting of a large number of molecules in the fluid state, the molecules are free to move within the boundaries of the system. For the case of real or solid boundaries, the moving molecules will strike the walls of the container and change direction. The acceleration resulting from the change of direction causes a force to be exerted on the wall of the container in accordance with Newton's laws of motion. If the fluid is assumed to be a continuum, a large number of molecules bombard the boundary and give rise to a continuous force against any reference area. Taking the smallest area over which a continuum condition can be assumed, the fluid pressure is defined as the normal force per unit area that exists on that small area of boundary. A fluid always has

TABLE 3.1
THERMOCOUPLE SPECIFICATIONS

Type	Tolerance Class 1	Tolerance Class 2
B	± 1.5°C (0°C to 375°C)	±0.25% (600°C to 1700°C)
E	± 1.5°C (–40°C to 375°C) ± 0.4% (375°C to 800°C)	± 2.5°C (–40°C to 333°C) ±0.75% (333°C to 900°C)
J	±1.5°C (–40°C to 375°C) ±0.4% (375°C to 750°C)	± 2.5°C (–40°C to 333°C) ± 0.75% (33°C to 750°C)
K	±1.5°C (–40°C to 375°C) ±0.4% (375°C to 1000°C)	± 2.5%°C (–40°C to 333°C) ± 0.75% (333°C to 1200°C)
N	± 1.5°C (–40°C to 375°C) ±0.4% (375°C to 1000°C)	± 2.5%°C (–40°C to 333°C) ± 0.75% (333°C to 1200°C)
R	±1.0°C (0°C to 1100°C) ±1.003% (1100°C to 1600°C)	±1.5°C (0°C to 600°C) ±0.25% (600°C to 1600°C)
S	±1.0°C (0°C to 1100°C) ±1.003% (1100°C to 1600°C)	±1.5°C (0°C to 600°C) ±0.25% (600°C to 1600°C)
T	±0.5°C (–40°C to 125°C) ±0.4% (125°C to 350°C)	±1.0°C (–40°C to 133°C) ± 0.75% (133°C to 350°C)

pressure. As a result of molecular collisions, every part of a fluid experiences forces on it, either by adjoining fluid or by an adjoining solid boundary. Pressure is a scalar quantity with S.I. units Newtons per square meter (N/m^2), Pascal (Pa).

The standard atmospheric pressure is taken as one atmosphere = 1.01325×10^5 Pa.

Note: 10^5 Pa = 1 bar
 1000 mbar = 1 bar
 100 Pa = 1 mbar

Absolute pressure is the pressure relative to vacuum.

Gauge pressure is the pressure recorded by a pressure-measuring device relative to atmospheric pressure.

Pressure Measuring Devices

Barometer

A barometer comprises a tube with one end closed. This initially is evacuated, and the open end is inverted and immersed in a reservoir of mercury, as shown in Figure 3.5. The reservoir is open to the atmosphere. Under standard conditions for pressure and temperature, the mercury will rise 760 mm above the reservoir.

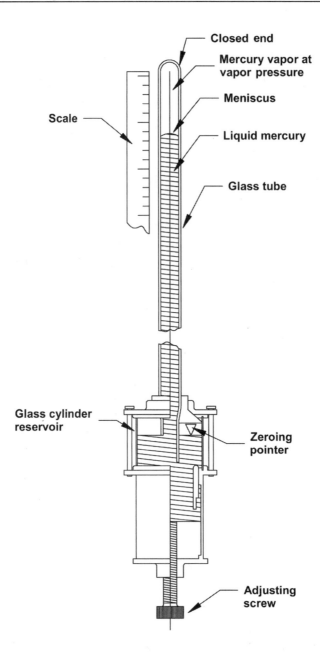

Figure 3.5 A barometer.

Manometer

A typical manometer consists of a transparent U-tube filled with a liquid such as water, alcohol, or mercury, as shown in Figure 3.6. If water is selected, a substance such as fluorescence can be added to the water to enable easier visualization. A single drop of detergent to 2 liters of water will change the surface tension and present a meniscus that is easy to read. This is a simple yet, when used intelligently, accurate device for monitoring steady-state non-pulsating pressures. Deflection of the surface level is proportional to the change of pressure. As the manifold pressure changes due to opening and closing of the throttle, the liquid moves up and down the scale. The manometer should be mounted as solidly as possible so an accurate reading can be obtained. Care should be taken not to subject the manometer to shock because the glass tubing is very fragile and is easily cracked. Calibration is performed by disconnecting the manometer and resetting the zero at atmospheric pressure.

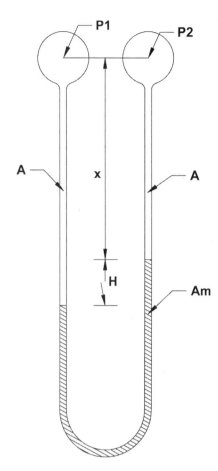

Figure 3.6 A U-tube manometer.

A variation of the U-tube manometer is the micro-manometer shown in Figure 3.7. These devices are used to measure very small pressure differences, down to as little as 0.05 Pa. In a micro-manometer, the reservoir is moved up or down until the level of the fluid within the reservoir is at the same level as the set-mark. Changes in pressure cause the fluid level to move, and the reservoir must be repositioned to be in line with

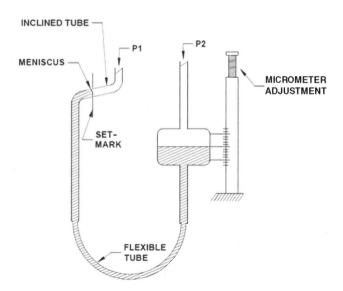

Figure 3.7 A micro-manometer.

the set-mark. The height change of this repositioning is equivalent to the pressure head due to the pressure difference.

The inclined tube manometer also is used to measure small differences in pressure (Figure 3.8). It provides increased sensitivity in comparison to a vertical manometer. The angle of inclination typically is between 10° and 30°.

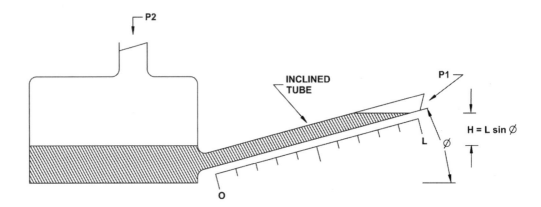

Figure 3.8 An inclined-tube manometer.

Transducers

A pressure transducer converts a measured pressure into a measurable quantity that may be mechanical or an electrical signal. The primary sensor usually is an elastic component that deforms or deflects under pressure. Several common types are illustrated in Figures 3.9 through 3.14, including the Bourdon tube, bellows, diaphragms, and capsules. Figure 3.9 shows pressure capsules.

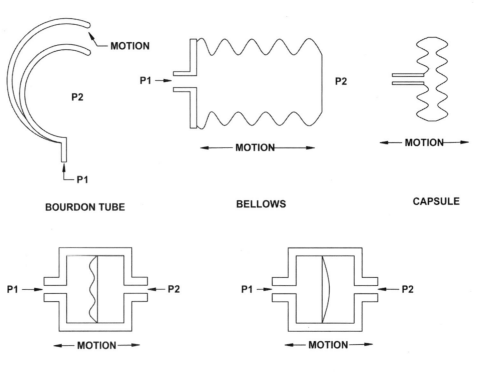

Figure 3.9 Pressure capsules.

The Bourdon tube (Figure 3.10) is a curved metal tube with an elliptical cross section that deforms with increasing pressure. One end of the tube is clamped, and the other end is connected to a mechanical linkage and indicator.

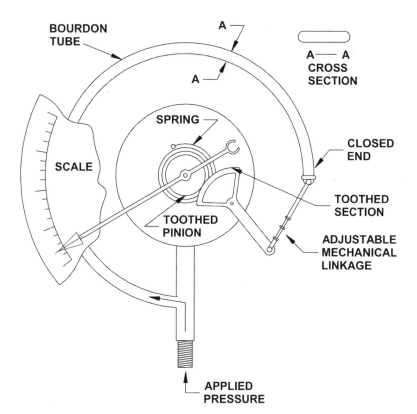

Figure 3.10 A Bourdon tube pressure gauge.

A bellows can be connected to a conductor, as shown in Figure 3.11, forming a potentiometric pressure transducer. Movement of the bellows causes a change in the potential difference indicated and therefore provides a measure of pressure.

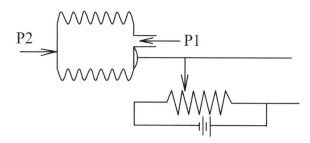

Figure 3.11 A potentiometric pressure transducer.

Diaphragm pressure transducers are a common choice. In these, a thin elastic circular disc or membrane is supported around its circumference. Under pressure, this disc will deform, and the amount of deformation can be measured by various means.

Diaphragm transducers are useful for measuring both static and unsteady pressures. The low mass and high stiffness of most membranes means that they usually have a very high natural frequency and a small damping ratio. The most common means of converting the diaphragm displacement into a measure of pressure is to sense the strain induced in the membrane, as illustrated in Figure 3.12.

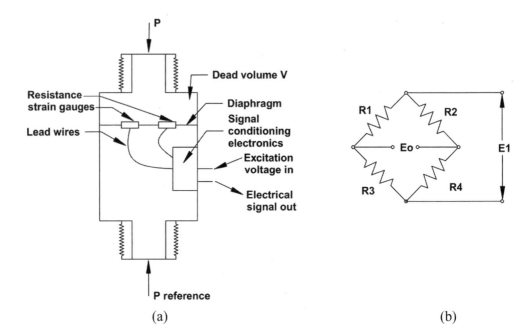

Figure 3.12 A diaphragm pressure transducer: (a) sensing scheme, and (b) bridge-strain gauge circuit for pressure diaphragms.

If a fixed plate is located above a metal or metalized diaphragm, then a measurement of the capacitance can provide an indication of the distance between the plates and therefore the pressure. Under the action of tension, compression, or shear, a piezoelectric crystal will deform and will develop a surface charge q, which is proportional to the force causing the deformation. Figures 3.13 and 3.14 show illustrations of piezoelectric pressure transducers.

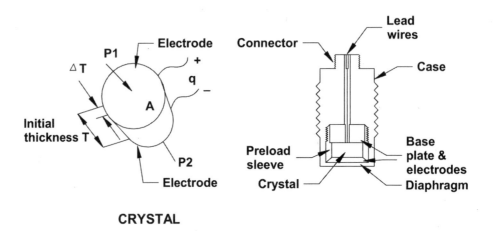

Figure 3.13 A piezoelectric pressure transducer.

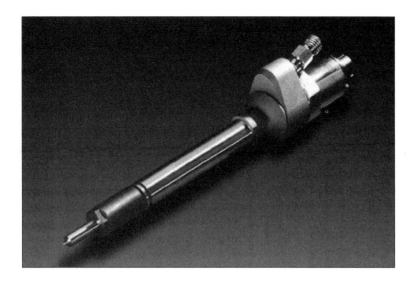

Fig 3.14 An example of a piezoelectric pressure transducer.

Pressure transducers and transmitters convert an applied pressure into an electrical signal. This signal is both linear and proportional to the applied pressure. Pressure sensors, transducers, and transmitters are commonly referred to simply as pressure transducers. The output electrical signal is sent to computers, chart recorders, digital panel meters, or other devices that interpret this signal and use it to display, record, and/or change the pressure in the system being monitored. The most popular output signal is a 4–20 mA two-wire current loop. Other voltage signals such as 0–10 V DC output also are used in some applications. All transducers require an input (also referred to as excitation or supply voltage) to power the internal circuitry. Wika transducers use two types of strain gauge sensors: piezoresistive, and thin film. Piezoresistive sensors are used in low-pressure applications up to 300 psi, whereas thin-film sensors are used above 300 psi.

Types of Pressure Transducers

General-purpose transducers provide excellent performance, reliability, and value for a wide variety of electronic pressure measurement applications. The accuracy is less than or equal to 0.5% full span. These units are available in pressure ranges 0–15 psig up to 0–5000 psig. General-purpose transducers are available with a 4–20 mA or 0–10 V DC output.

Industrial-grade transducers are precision engineered to fit most industrial pressure measurement applications. Each unit undergoes quality control testing and calibration to achieve an accuracy of 0.25% full scale. The printed circuit board uses state-of-the-art surface mount technology and is potted in silicone gel for protection against vibration, shock, and humidity. Industrial models have a 4–20 mA output signal and offer pressure ranges including vacuum, compound, and pressures up to 5000 psig. Advantages of the industrial-grade model include higher accuracy, zero and span adjustment for recalibration, and better resistance to vibration, shock, and humidity than most transducers on the market.

Pressure transducers come in various sizes, depending on the magnitude of the pressure to be measured. The transducers usually are found inside the "boom box" within the cell. Fittings are drilled and tapped into the engine in the appropriate place and are connected to the boom box via a pipe. The pressure is converted into a voltage signal by the transducer. This voltage can be shown as readout on the screen by the data-logger. Calibration is performed in-house by feeding the transducer with an exact pressure and then comparing this to the readout given on the screen. If inaccurate, a correction figure can be entered. Figure 3.15 illustrates the installation of a pressure transducer.

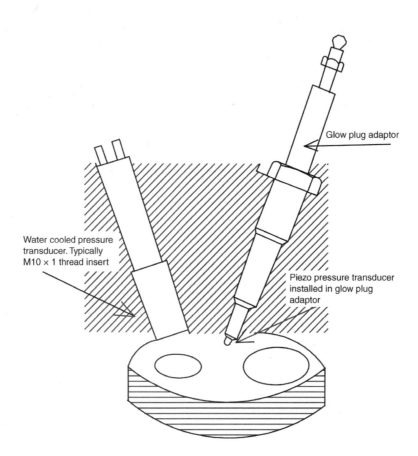

Figure 3.15 *Installation of in-cylinder pressure transducers.*

Flow Measurement

The purpose of flow measurement is to measure the flow of fluids around the engine. The most common application is for measuring coolant and/or oil flows within the engine. Coolant flow is a requirement if a thermal balance test series is to be undertaken.

Several methods and devices are used in the measurement of flow, such as the venturi gas meter (Figure 3.16) and the flow turbine (Figure 3.17). The venturi meter measures flow rate in terms of pressure drop across a venturi (or tapered throat) within a pipe. The turbine is fitted directly into a pipe to measure the flow rate of the fluid. The operating principle is that the flow impinges on the turbine blades, causing them to rotate. The rate of rotation is measured either mechanically or electrically. In the latter case, it works by giving off an electrical output directly related to the speed of the turbine, which, in turn, is directly related to the rate of flow. The electrical output then is measured using a frequency counter that can easily be converted into a flow figure (i.e., liters per minute). The manufacturer performs the calibration.

Figure 3.16 Venturi gas meter. (Courtesy of University of Sussex)

Figure 3.17 Turbine flow meter. (Courtesy of University of Sussex)

Mass Airflow Sensors

This type of sensor (Figure 3.18) is used in the Bosch K type Jetronic fuel injection system and the RDA mass airflow system (CMC-Scotch Yoke three-cylinder application). It consists of an air funnel and a pivoting airflow sensor plate. A counterweight compensates for the weight of the sensor plate and the pivot assembly.

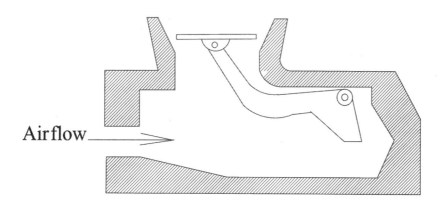

Figure 3.18 Mass airflow measurement.

Rotameters

A basic rotameter consists of a small float supported in a tapered glass tube by a flow of liquid or gas. In engine flow measurement, blow-by gases are directed through the calibrated tapered tube. Inside the tube, a calibrated pintle is held in suspension by the blow-by gases. The distance up the tube is proportional to the gas flow, thereby indicating the rate of flow. Figure 3.19 shows an example of a rotameter.

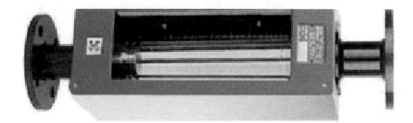

Figure 3.19 Typical glass-walled rotameter. (Courtesy of University of Sussex)

Square Edged Orifice Plates

This is the preferred method of airflow measurement. The blow-by gases are directed into a parallel round tube. Positioned within the tube is a disc with a regular hole in it. Pressure measurements are taken on either side of the orifice. By measuring the pressure difference across the orifice, it is possible to calculate the gas flow. (See British Standard 1024 and Figure 3.20.)

Figure 3.20 British Standard square edged orifice plate.

With reference to Figure 3.21, D (internal diameter of pipe) and 0.5D tappings measure the pressure difference between one pipe diameter upstream and a half pipe diameter downstream from the orifice plate. The presence of the orifice within a pipeline will cause a static pressure difference between the upstream side and the downstream side of the device.

The installation and use of orifice plates are documented in ISO 5167-1:1991(E) (also BS1042: Section 1.1:1992). The information given here has been sourced from this document and refers only to D and 0.5D pressure tappings.

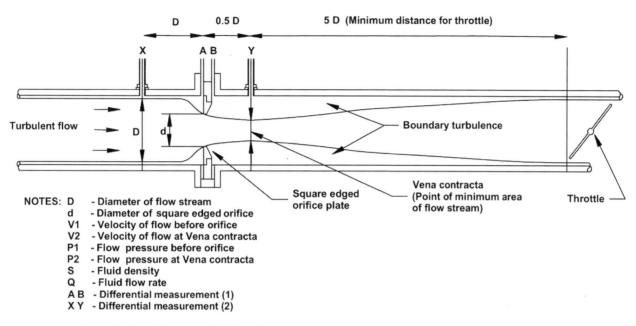

Figure 3.21 Standard airflow installed orifice.

The mass flow rate through an orifice plate installation can be determined from the following equation:

$$q_m = \frac{C}{\sqrt{1-\beta^4}} \varepsilon_1 \frac{\pi}{4} d^2 \sqrt{2\Delta p \rho_1}$$

where

C = coefficient of discharge

β = diameter ratio = d/D

d = diameter of orifice (d)

D = upstream internal pipe diameter (d)

ε_1 = expansibility factor

Δp = differential pressure (Pa)

ρ_1 = density of the fluid (kg/m³)

q_m = mass flow rate (kg/s)

The density can be evaluated from conditions at the upstream pressure tapping ($\rho_1 = P_1/RT$). The temperature should be measured downstream of the orifice plate at a distance of between 5D and 15D. The temperature of fluid upstream and downstream of the orifice plate is assumed constant (Figure 3.22).

56 An Introduction to Engine Testing and Development

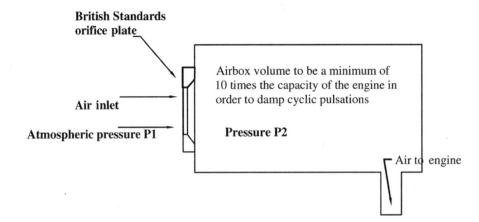

Figure 3.22 Base airflow measurement system, 45-gallon fuel drum.

Lucas-Dawe Air Mass Flow Meters

This flow meter originally was intended for use with engine management systems. (These were superseded by the whetstone bridge hot wire systems.) However, this meter is better suited to laboratory use. The principle, as illustrated in Figure 3.23, shows the central electrode that is maintained at approximately 10 kV so that a corona discharge is formed. The exact voltage is varied so that the sum of the currents flowing to the two collector electrodes is constant. When air flows through the duct, the ion flow is deflected, thereby causing an imbalance in the current flowing to the two collector electrodes. The difference in current flow is proportional to the mass airflow rate.

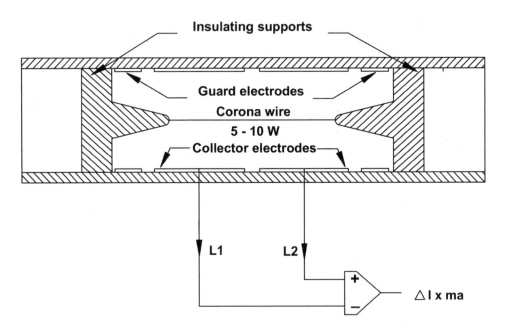

Figure 3.23 A Lucas-Dawe air mass flow meter.

Calibration

Calibration is all, regardless of the objectives of the test in hand. All key instrumentation readouts must be within defined calibration limits. Failure to calibrate could mean that the test results are questionable, with the test series being a waste of valuable time and money. The calibration principles detailed in this chapter are based on the general requirements of European Quality Assurance Standards EN45001 and EN ISO 9000.

The calibration process can be defined as the set of operations, which, under controlled conditions, establishes the relationship between values reported by the target measuring chain and the corresponding known values measured via a calibration transfer standard. The calibration process should target a number of calibrated points that are representative of real values anticipated during normal operating conditions:

- To calibrate prior to use, and to adjust at prescribed intervals, the measuring and test equipment against certified equipment having a known valid relationship to nationally recognized standards.

- To establish and maintain procedures to ensure that non-conforming equipment is prevented from inadvertent use or installation.

Calibration for all elements critical to the utility of the test should always be traceable to national standards. The frequency of calibration should be determined by a scientific approach, paying due attention to the past behavior of the measurement system used. Evidence of calibration should be readily discerned on all measurement equipment. Wherever possible, the calibration of each parameter should be performed with reference to the whole measurement system. Calibration should include all instrumentation between the physical quantity being measured and the point at which data are logged for use with the test report. For example, calibration of a temperature sensor would require the sensor to be placed in a temperature-controlled bath and the measurement recorded employing the instrumentation used during the test. Alternatively, persuasive evidence should be provided for using other direct methods where these are considered appropriate.

Where possible, all critical equipment shall be tagged or labeled with the serial number, frequency of calibration, and calibration status shown on the tag or label.

Calibration of each parameter and its associated measuring chain must be performed at regular intervals not exceeding 12 months. On completion of calibration, the date of the calibration should be defined and the date documented. Regular calibration intervals should be defined, based on the history of each parameter and the associated measuring chain. If any parameter/measuring chain is suspected to be out of calibration, then an intermediate calibration check must be performed. Similarly, if any component in the measuring chain is replaced, then an intermediate calibration check must be performed. Any excursions outside the permitted limit of error for that parameter must be documented, with the analyzed reason for the unexpected occurrence recorded.

Definitions of Calibration Terms

- **Operational range**—This is defined as the widest possible range of values (difference between the maximum and minimum values) anticipated to be seen for any given parameter throughout the duration of the test.

- **Parameter/measuring chain**—All items of hardware with the potential to affect the reported value in engineering units must be identified. Examples include a sensor, cable, signal conditioning, and display/logging resolution.

- **Parameter calibration range**—This is the maximum recommended range over which each specific parameter should be calibrated.

- **Calibration range**—The range (minimum to maximum) over which each specific parameter/measuring chain is calibrated should exceed the operating range.

- **Full-scale deflection (FSD)**—The maximum single-sided values for each measuring chain having a calibrated range intercepting at or through zero.

- **Limits of error (LOE)**—The maximum plus or minus error acceptable with reference to the calibration transfer standard, or the defined limits of error excluding uncertainty errors associated with the calibration transfer standard and method of calibration.

- **Accuracy of measurement**—The closeness of the agreement between the result of a measurement and the true value of the measurement.

- **Uncertainty of measurement**—Results of the evaluation are aimed at characterizing the range within which a true value of measurement is estimated to lie, generally with a given likelihood.

- **Traceability**—The property of the result of a measurement whereby it can be related to appropriate measurement standards (generally international or national), through an unbroken chain of comparison.

Calibration Personnel and Equipment

All personnel carrying out calibration should be adequately trained and regularly reviewed to assess their competence with the calibration equipment and procedures. All the channels defined as critical calibration parameters must be calibrated using equipment that is fit for that purpose and traceable to national standards. Equipment records must be kept, detailing the history of all calibration standards used.

A pressure transducer can be calibrated by direct comparison against a reference pressure instrument or by subjecting the transducer to a verifiable pressure using a dead-weight tester (Figures 3.24 and 3.25). Dead-weight testers comprise a piston, a reservoir, and a means to connect a pressure-sensing device to the reservoir. In the operation of some devices, the piston must be spun to indicate that it is riding on the reservoir liquid as opposed to resting on the supports. These devices are useful for calibration of pressure sensors in the range of 70 Pa to 700 MPa. The frequency response and rise time of a pressure transducer must be found by dynamic calibration.

Temperature Calibration

Calibration can be taken to mean the establishment of the relationship between the transducer output and the transducer temperature within a tolerance or band of uncertainty. Calibrations can be considered to fall into one of four categories:

1. Acceptance tests
2. Batch calibration
3. Calibration of a single thermocouple for a fixed application
4. Calibration of reference standards thermocouples

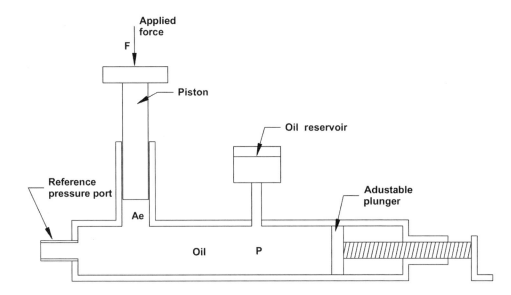

Figure 3.24 A dead weight tester.

Figure 3.25 A commercial dead-weight tester. (Courtesy of University of Sussex)

Chapter 4

An Introduction to Mr. Diesel and His Engines

The original patent for the principle we now refer to as diesel was taken out by Charles Stewart. However, it was Rudolf Diesel who took it up and developed the concept into a mobile practical device.

Rudolf Diesel, a nationalized German engineer, was born in Paris, the son of a leather worker. In the 1860s under Emperor Napoleon III, war with the non-unified German states made things very difficult for the Diesel family. Thus, they moved to England, where they set up home in Newhaven. In 1870, they moved to London, where Rudolf spent much of his spare time in the science museum (then the South Kensington Museum), studying the science exhibits. This laid the foundations for his future work as an engineer. Rudolf Diesel made and lost several fortunes, and he died in a mysterious manner. He disappeared from the ferryboat, *Dresden*, between Antwerp and Harwich, and left an empty bank account, no reserve cash, and mounting debts.

When he built the first functional diesel engine in 1897, Rudolf Diesel certainly was not able to foresee what would become of his invention. Since then, the diesel engine has gone through a number of decisive advances in design. Particularly in the last few years, the diesel engine has become even more attractive because of its low fuel consumption and several important developments in the diesel sector, such as common rail, high-pressure injection, and piezo injectors. Unlike the gasoline engine, the diesel engine operates without an ignition spark. However, the engine is constantly evolving, and the very low emission levels that are being considered by international legislators may mean that compression ignition is not accurate enough. Indeterminate valve seating, and uncontrolled internal exhaust gas recirculation due to valve overlap, may force the onset of spark ignition flame propagation. In the conventional compression ignition diesel engine, the fuel is combusted due to auto-ignition. Combustion takes place in a cylinder. A piston compresses the air drawn into the cylinder so that its temperature rises sharply. The fuel is introduced (injected) into the hot air and combusts almost completely. The combustion causes the piston to move downward, so that the crankshaft is turned and the engine runs.

Early in his development work, Rudolf Diesel discovered that high pressures were necessary for injection of the fuel. In the early stages, this high pressure was generated by an air compressor, which was not only very heavy but also costly. In the 1920s, Robert Bosch developed the injection pump and so started the Bosch empire. However, many others led the way, such as Knut Hesselman (1908) and James McKechnie (1910).

When Rudolf Diesel contacted Augsburg and Krupp of Germany in 1893 to develop a more efficient internal combustion engine, one of his key objectives was to use powdered coal as a fuel. Mountainous piles of powdered coal were accumulating in the German countryside along the Ruhr valley; thus, ready supplies were available.

The first experimental coal-dust-burning engine was built and run in 1893 and used air to blast the fuel into the combustion chamber (Figure 4.1). It is interesting to note that the same principle is used today in the air blast fuel injection system produced by the Orbital Corporation Ltd. In Mr. Diesel's first experiments with oil, it was mechanically injected into the engine. The results were poor due in part to the crude injection equipment and the consequent large volume of dead fuel. When he returned to his air blast system, then burning oil was very successful and became the accepted method for many years.

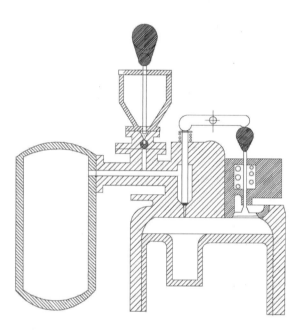

Figure 4.1 Air blast coal dust delivery system.

In 1908, a Swedish engineer named Knut Hesselman invented a small injector using the air blast principle. The metered oil was deposited in an annular space above the valve seat. When the valve was lifted, an aspirating effect was produced on the oil to discharge it along with the air flowing past into the combustion chamber. Early Krupp engines used single-stage compressors with little commercial success. Three-stage compressors built by the Diesel Motor Company of America enabled the system to work very efficiently for its day.

In 1910, James McKechnie from the Vicars Company in England produced the first airless injection system (Figure 4.2). In this system, a metering pump delivered oil to a spring-loaded plunger, which was raised by a cam. Tripping of the cam allowed fuel to be injected into the cylinder of the engine as the spring returned the plunger to its bottom position.

In 1913, Vicars developed a common rail system that became very popular. A multiplunger pump delivered fuel to an accumulator with the fuel pressure maintained at 5000 psi by means of a relief valve. The fuel was sprayed into the engine cylinders through mechanically operated injection nozzles. The duration of the plunger stroke

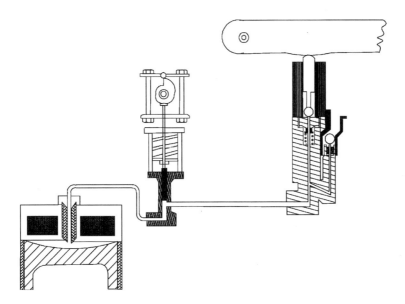

Figure 4.2 The McKechnie non-air blast system.

during which the bypass valve, D, was closed was lengthened or reduced by moving the wedge, K, in or out, thus increasing or decreasing the fuel quantity delivered by the pump.

A major development was Rudolf Diesel's metering pump (Figure 4.3).

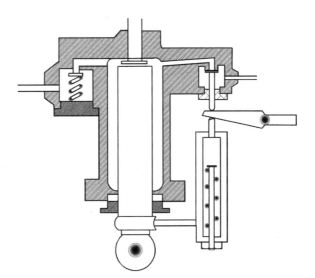

Figure 4.3 Rudolf Diesel's metering pump.

In 1914, a Belgian called Francois Feyens invented the distributor rotary pump (Figure 4.4), and in 1910, Peter Bowman of Denmark invented the pintle valve as we know it (Figure 4.5). Hans Heinrich of Robert Bosch took up the latter patent in 1935, and world patents were granted to Bosch, thus establishing the popular throttling pintle nozzle (Figure 4.6). The pintle, or valve extension, protruded through the spray hole to produce an annular orifice, and the seat for the flat-bottomed valve was close to the orifice so that the nozzle had a differential action.

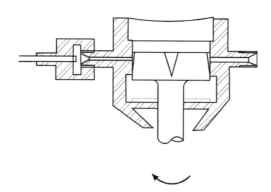

Figure 4.4 Distributor rotary pump.

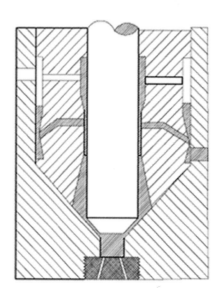

Figure 4.5 Early pintle nozzle.

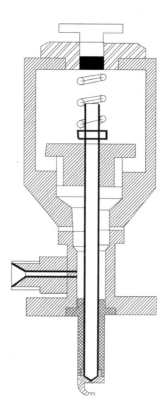

Figure 4.6 Inwardly opening pintle nozzle.

Unit Injectors

From day one in the history of diesel-fueled engines, troubles were encountered with the fuel discharge tubing connecting the injection pump to the nozzles. One method of overcoming this was to combine the pump and the nozzle.

Carl Wiedman of Germany patented the plunger system (Figure 4.7) in 1905. The fuel quantity delivered by the plunger, and controlled by the opening period of the suction valve, passes through the check valve into the atomizer. There it is mixed with pressurized air before ejection through the nozzle orifice.

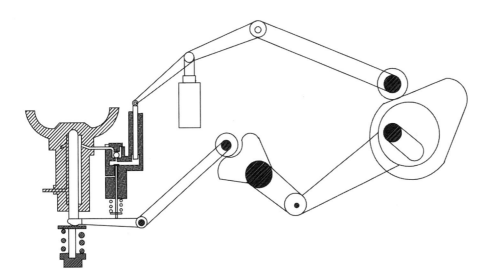

Figure 4.7 Plunger system.

Advantages and Shortcomings of Diesel Engines

Prior to studying the various diesel fuel injection systems, it is worth considering the advantages and shortcomings of the diesel engine as we understand it today.

Disadvantages of Diesel Engines

Normally, diesel engines have a weight per horsepower that is approximately 1.5 to 3 times that of a gasoline (petrol) engine. Why? The reason for this increased weight is that a diesel engine relies on self-ignition, which depends on high-temperature compressed air induced by a high compression ratio. As a result, its combustion pressure (P Max) is much higher than that of a gasoline engine.

To withstand the high pressure and rapid rate of pressure rise, a very strong structure is required. The combustion pressure of a diesel engine is approximately 1.5 times higher than an equivalent capacity, normally aspirated gasoline engine. When the swept volume is the same, the diesel engine produces only about two-thirds the horsepower of the gasoline engine.

So, why is the power output smaller, even though the combustion pressure is higher? The answer is as follows. Because combustion is an oxidation reaction, there is a specific weight of air that will completely oxidize one gram of fuel without leaving excess oxygen. This weight of air is called the stoichiometric air/fuel ratio. The gasoline engine operates with an air/fuel mixture that is very near the stoichiometric air/fuel ratio. It is necessary to have this ratio because an air/fuel ratio much greater than stoichiometric is difficult to ignite with a spark plug, and one much smaller than stoichiometric (fuel rich) is inefficient. In the diesel engine, the fuel is injected into the combustion chamber at a point near the end of the compression stroke, and the fuel ignites spontaneously. As mixing occurs between the fuel and air, burning continues. This process is extremely heterogeneous. Soot (black smoke) is formed during combustion because some of this fuel has insufficient oxygen for complete combustion. As more fuel is injected, more and more soot is produced. Hence, the air/fuel ratio of the diesel engine must always be higher than stoichiometric to prevent excessive amounts of soot. To reduce the amount of soot means that less fuel is present in the diesel engine cylinder than in a cylinder of an equivalent gasoline engine, and the diesel engine power therefore is reduced. For the same swept volume, current diesel engines can use only 70 to 80% of the fuel used by a gasoline engine. (This is not the case in turbocharged applications.) In diesel engines, the utilization factor varies with the combustion system/fuel delivery system. For example, a diesel engine with a pre-combustion chamber system (indirect injection [IDI]) has a utilization value of approximately 80%, whereas a normally aspirated direct injection (DI) system, which is used mainly for large trucks, has values of approximately 50 to 70%. Direct injection systems still are not as well developed as IDI pre-combustion systems, but developments are moving forward at a rapid pace.

Another reason for the output of a diesel engine being less than that of a gasoline equivalent swept volume engine is that gasoline engines can operate at high speeds because the combustion rate increases with the engine speed. On the other hand, diesel has a reduced combustion efficiency at high speeds because of the longer ignition delay, slow mixing of the mixture, and longer injection duration (in terms of crank angle). Note that this is not as relevant with electronic diesel injectors; those based on the piezo crystal expansion with current have infinitely fast response. The international governmental smoke limits are difficult to meet at high speeds. The diesel engine has a very high compression ratio, and the energy required to rotate the crankshaft (i.e., friction horsepower) is larger than that of the gasoline engine. The friction loss is large in proportion to the engine rotational speed. Thus, when the speed is increased to boost the output, the friction losses increase and thereby negate the increase in output. The maximum speed that a current Formula 1 engine can reach is 17,000 rev/min; a diesel can reach only approximately 5000 rev/min.

Advantages of Diesel Engines

We have reviewed the many disadvantages of diesel engines, so one might wonder why the diesel engine is used? The answer is that a diesel engine has very good fuel consumption; indeed, it is so good that it offsets the many disadvantages.

When comparing fuel consumption per horsepower per hour at the maximum output between the best diesel and the best gasoline engines of similar swept volumes and valve timing configurations, it will be found that the diesel engine uses only approximately 70% of the fuel that the gasoline engine uses. In part-load condition—the condition in which on-road application is normal automobile driving, idle to half throttle—the fuel consumption of a diesel engine can be as little as 60% of that of the gasoline engine.

Why Does the Diesel Engine Have Such Good Fuel Consumption?

The good fuel consumption of the diesel engine is due to the high compression ratio required for self-ignition and the reduced pumping losses because the diesel has no induction air throttle plate. The higher the compression ratio, the better the thermal efficiency. The gasoline engine cannot accommodate high compression ratios because of the onset of the knocking phenomenon.

The theoretical thermal efficiency is given by

$$\eta^{th} = 1 - \frac{Q_L}{Q_h} = 1 - \frac{1}{\varepsilon^{k-1}} = 1 - \frac{T_1}{T_2} = 1 - \frac{T_4}{T_3}$$

where ε = ratio of compression and expansion (compression ratio)
  Q = ideal gas
  k = specific heat at a constant pressure/specific heat at a constant volume

That is, the theoretical thermal efficiency is related only to the compression ratio and K, and it corresponds to the efficiency of the Carnot cycle active between temperatures T_1 and T_2 or T_3 and T_4.

As long as fuel economy is measured using fuel volume, diesel fuel has another advantage. Diesel fuel has a specific gravity that is approximately 10% higher than that of gasoline, that is, one liter of diesel fuel is 10% heavier than one liter of gasoline. The amount of energy in a specific weight of diesel fuel thus is 10% greater than that of gasoline. When fuel economy is measured by volume (i.e., kilometers per liter or miles per gallon), diesel fuel would produce 10% greater fuel consumption than gasoline, even if the engines otherwise were identical.

Diesel uses an air/fuel mixture with approximately 40% excess air; the gasoline engine runs with a stoichiometric mixture. The leaner mixture results in higher efficiency because with a lean mixture, the combustion temperatures are lower with sufficient oxygen. Lower combustion temperatures reduce heat loss.

A better fuel consumption signifies that the energy in the fuel is used efficiently and at a lower combustion temperature due to a higher air/fuel ratio, which reduces the heat quantity released to the coolant. This reduced heat means that the radiator can be downsized, and the cooling fan can be reduced in size. In some gasoline-to-diesel conversions, the radiator can be reduced by more than 35%.

A key advantage of the diesel engine is its durability, resulting in a longer service life as a direct result of the robust and heavy structure required to sustain the high combustion pressures.

As we have discussed, the peak cylinder pressure of the diesel engine is more than 1.5 times that of the gasoline equivalent. If the same stress is put on a structure but at a higher pressure, then that structure must be thicker. For example, consider a round bar with a diameter D loaded as shown in Figure 4.8.

The stress is proportional to $1/D^3$, while the deflection is proportional to $1/D^4$. Namely, to keep the stress the same, the increase in the diameter must have a greater rigidity that is proportional to the increase of the diameter. If a diesel engine is designed for

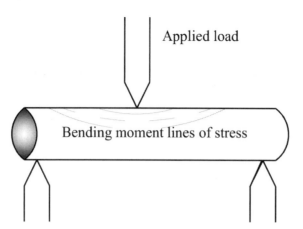

Figure 4.8 Bending moment.

a higher cylinder pressure, its deflection would be similar to that of a gasoline engine designed for a lower cylinder pressure. However, note that higher crankcase rigidity always gives better durability.

Fuel Injection and Combustion Principles

In December 1922, Robert Bosch of Stuttgart decided to manufacture fuel injection equipment. In a relatively short period of time, other companies started manufacturing standardized types of diesel fuel injection equipment for all types of engines, and progress in diesel engine development swung into high gear. To be totally confident that we have a full and complete understanding of the combustion principle, let us again examine each of the combustion principles and compare the different combustion systems.

Pre-Chamber Systems

In the pre-chamber system for passenger car diesel engines, the fuel is injected into a hot pre-chamber. Pre-combustion is initiated to achieve good mixture formation with reduced ignition lag for the main combustion process. The fuel is injected via a throttling pintle nozzle at a low pressure (circa 300 bar). A specially designed baffle surface in the center of the chamber distributes the jet of fuel when that jet hits it, causing further atomization, and mixes it with the incoming air (Figure 4.9).

Combustion starts and drives the partially combusted air/fuel mixture through bores at the bottom end of the pre-chamber into the main combustion chamber above the piston, with the mixture gaining heat in the process. Here, extensive mixing takes place with the air in the main combustion chamber.

A strong air vortex is generated during the compression stroke, and the fuel is injected into this swirling air to give better mixing. The nozzle is placed so that the jet of fuel hits the swirling air at 90° to its axis and hits the opposite chamber wall in a hot zone. Compared with the pre-chamber process, the flow losses between the main and auxiliary combustion chambers are lower with the whirl-chamber system (Figure 4.10) because the throat cross-sectional area is greater, thus giving lower cycle-to-cycle pumping losses with, as a consequence, greater efficiency and fuel consumption advantages. The glow plug position is critical and must be matched carefully to ensure good mixture formation at all speeds and load conditions.

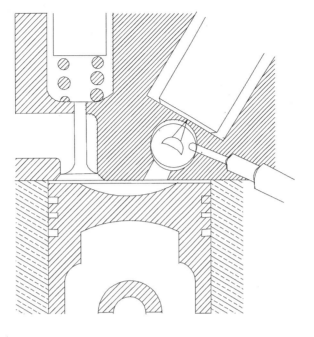

Figure 4.9 Pre-chamber with baffle.

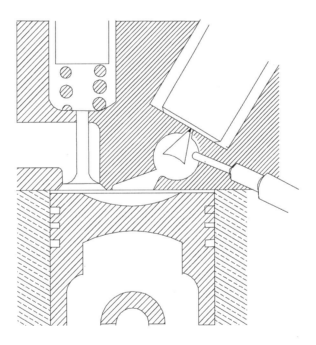

Figure 4.10 Whirl-chamber process.

The position of the glow plug is always a compromise between the ease of starting and the fuel mixture under load. A key element in worldwide current legislated emission regulations is the need for rapid heating of the whirl chamber after cold start. This heating reduces ignition lag and avoids the production of unburned hydrocarbons (HC) in the exhaust during warm-up. Figure 4.11 is an illustration of a glow plug.

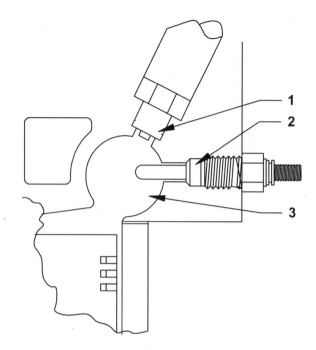

Figure 4.11 Sheathed-element glow plug. (1) Injector. (2) Sheathed element glow plug. (3) Swirl combustion chamber.

Direct Injection Systems

In the past, direct injection (DI) systems have been used mainly in commercial vehicle and stationary off-highway diesel engine applications. The mixture formation takes place in a combustion chamber machined into the piston crown, as shown in Figure 4.12.

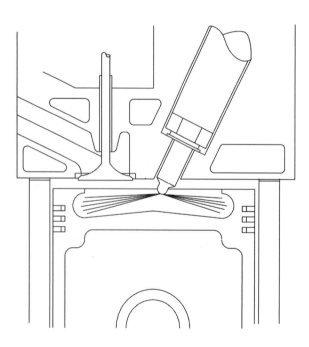

Figure 4.12 Direct injection.

For good mixing, a vortex is caused by the special shape of the intake port in the cylinder head. The design of the piston crown with its integrated combustion chamber contributes to the air movement at the end of the compression stroke (i.e., at the start of injection). In these applications, unlike the pre-chamber engine with a single jet throttling pintle nozzle, a multi-hole nozzle is used for the direct injection process. The spray position must be optimized in accord with the combustion chamber design.

There are two key methods of direct injection:

1. Mixture formation by controlled air movement

2. Mixture motion introduced by fuel injection without controlled air movement

In the latter case, a high number of very small injector holes feed at very high pressures. In the direct injection system previously described, mixture formation is achieved by the mixing and the evaporation of fuel particles with the air particles surrounding them.

A direct injection system with wall distribution is a system for commercial and stationary engines. In this system, the heat content of the piston-recess wall is used for evaporation of the fuel, and the air/fuel mixture is produced by suitable guidance of the combustion air (Figure 4.13). The process operates with a single-hole nozzle with a relatively low injection pressure. If, following extensive development, the air movement is adjusted correctly, extremely homogeneous air/fuel mixtures can be obtained with long combustion duration, low pressure rise, and quiet combustion. This is offset with increased fuel combustion when compared with air distribution (high swirl ports).

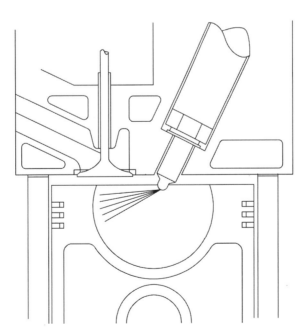

Figure 4.13 Direct injection with wall distribution.

Comparison of the Systems

The disadvantages of pre-chamber engines with regard to noise are most noticeable during cold running. Poor mixture formation caused by heat dissipation to the combustion chamber walls gives rise to long ignition lags and to knocking combustion noise. Under warm-up, the whirl-chamber engine also leads to high combustion noise in low-load and low-speed ranges.

The main advantage of direct injection systems is a reduction in fuel consumption of up to 20% compared with chamber engines.

To summarize, direct injection systems always require higher injection pressures and hence a more complex injection system.

An Introduction to Diesel Fuels

Diesel fuels consist of a number of hydrocarbons, which have boiling points in the range between 180 and 360°C.

Cetane Number

Because there is no external ignition system, diesel fuel must ignite when introduced into heated compressed air with the minimum possible delay. Ignition quality of the fuel is defined as the property of the fuel that serves to initiate auto-ignition. This ignition quality is expressed by the cetane number (CN). The higher the CN, the easier it is for the fuel to ignite. The numbers allocated range from 100 for very good, down to methylnaphthalene, which has very poor ignition qualities and a CN of 0.

The minimum CN for diesel fuel is 45; a CN of 50 is the optimum for current engines.

Cold Behavior

At low temperatures, paraffin crystals can cause filters to clog and block delivery of the fuel. For untreated fuel, this crystallization starts at 0°C. With this in view, cold weather additives are added at the refinery, and standard fuels are good to –22°C.

Density

The calorific value of diesel fuel is approximately related to its density. That it, it increases with increasing density. Thus, if fuels with differing density are used with the same fuel injection equipment at the same settings, there will be fluctuations in calorific values, which in turn will lead to increased smoke and soot emissions from higher-density fuel.

Calorific Value

The calorific value of diesel fuel is lower than that of gasoline. Diesel fuel ranges from 40 to 43 Mj/kg and is a function of fuel density.

Note that the higher the sulfur content, the lower the calorific value. This has a big impact on new emission regulations.

The Diesel Engine and International Regulations

The diesel engine is now as popular as the gasoline-fueled engine, but ever more stringent international governmental emission regulations require advanced after-treatment work. There are a number of options, including the following.

The nitrogen oxides (NOx) trap has the following advantages:

- Provides NOx reduction of ≈50% on the NEDC [North East Diesel Collaborative (emission reduction)]

- Does not require additional reductant

- A good potential environmental image

The NOx trap presents the following issues:

- Sulfur regeneration required (650°C rich, 10–15 min, 10^3-km intervals)

- NOx regeneration required (rich, 3–5 sec, ≈100-sec intervals)

- Limited load/speed range for rich operation

- Lean–rich–lean transitions must be transparent to the driver (torque, noise, smoke)

- Limited operating window (200–450°C)

- Optimum NOx trap design (chemistry) may not provide fast light-off for HC and CO

- Fuel economy penalty (currently 5% on NEDC)

- Fuel sulfur requirements less than 10 ppm, ideally 0 ppm (to minimize SOx regeneration)

- Risk of HC and CO breakthrough during NOx regeneration

- Net effects of HC and CO increases during SOx regeneration are averaged over n-cycles

- Negative impact on trap durability due to SOx regeneration temperature

- Possible hydrogen sulfide (H_2S) emissions

- Electric onboard diagnostics (EOBD) to be developed

- Requires fully flexible fuel injection electronics (FIE) (i.e., common rail)

- High system cost

Diesel Particulate Filters

The diesel particulate filter (DPF) has the following advantages:

- Provides particulate emissions reduction of more than 90%

- No visible smoke

- Under suitable operating conditions, will "self generate"
- Enables the re-optimization of calibration for NOx control
- Positive environmental image

Diesel particulate filter has the following issues:

- Particulate regeneration is required (450°C, 10–15 mins, ≈500-km intervals)
- An additive system requires:
  - Additive introduction system
  - Servicing
- A coated system requires increased regeneration times
- Complex control and monitoring system
- Fuel economy penalty (currently 3% on NEDC)
- Engine onboard diagnostic (EOBD) system to be developed, with consideration of the following key issues:
  - Packaging
  - Cost
  - Weight
  - Suppliers

Diesel particulate filter and NOx trap combinations have the advantage of providing the benefits of a DPF and a NOx trap.

However, DPF and NOx trap combinations have the following issues:

- A diesel particulate filter and NOx trap compete for space and position in the exhaust system package
- The NOx trap must be ahead of the DPF to provide correct temperature conditions for effective NOx storage and SOx regeneration
- Lack of NOx at the DPF inlet prevents any continuous regeneration of the DPF and increases the filter regeneration frequency
- The NOx trap regeneration (rich) increases particulate deposition on the DPF and may adversely affect the particulate chemistry
- During DPF regeneration (lean), NOx trap regeneration is not possible, and this may lead to deep storage of NOx, making subsequent NOx trap regeneration more difficult
- It may not be possible to optimize the NOx trap as a DPF regeneration catalyst
- NOx trap sulfur regeneration may lead to sulfate issues with DPF and the potential for the generation of white smoke

Advantages of the continuously regenerating trap (CRT) filter are as follows:

- Provides particulate emissions reduction of more than 90%
- Provides continuous regeneration of trapped particulates (when conditions allow)

Continuously Regenerating Trap Filters

The following are issues related to the continuously regenerating trap filter:

- Uses NO_2 to oxidize trapped particulate. The NO_2-to-particulate ratio requirement is higher than normally achieved in European automotive diesel applications.

- If a vehicle is operated continuously in a low NOx mode, filter regeneration will not occur, and the system will store high levels of particulates. Potential consequences are as follows:

  - Increased backpressure stalls the engine

  - A loss of control when regeneration conditions are re-established, leading to filter failure

- Requires a very active oxidization catalyst ahead of the particulate trap to oxidize NO to NO_2. The NO to NO_2 reaction is inhibited by sulfur, with less than 10 ppm sulfur required.

- The highly active oxidization catalyst will turn sulfur into sulfate, which would be measured as particulate and could be released as white smoke.

- Fuel economy penalty due to backpressure increase (estimated as 3% on NEDC).

- Must be sited ahead of a NOx trap (if used), inhibiting its effectiveness on the NEDC (temperature too low) and making SOx regeneration difficult (650°C).

The diesel engine is a highly sophisticated machine, but like other engines, it has advantages and disadvantages. International legislation makes ever more demands on the automotive development engineer. As always, he or she is faced with many options, and compromises are always required. It is never black or white.

The principle of the compression ignition engine has been around since Charles Stewart took out the original patent in the nineteenth century, but it was Rudolph Diesel who went on to develop the potential of the engine. Throughout the twentieth century, progress with the diesel engine advanced, and it evolved into the familiar product that competes with the gasoline engine. The twenty-first century sees growing concern for global warming, which has been gathering for several decades; the emphasis is very much on reducing harmful emissions, so research on the diesel engine is likely to continue.

Chapter 5

Engine Tests Used Within the Automotive Testing Industry

This chapter deals with engine testing, in particular the durability tests used within the industry. Relevant tests and their applications are described, and detailed accounts are given of the processes and how to read and record the data. Examples are given of a typical test procedure and a master service record sheet. The remainder of the chapter documents the procedures of an engine test, from inception to completion.

Types of Tests

Many different types of engine tests are performed within the industry, some more common than others. The six principal tests and relevant applications are listed as follows:

1. **Durability (Design Validation Test)**
    - Steady load and speed operation
    - Load cycling
    - Speed cycling
    - Thermal shock cycling
    - Component development
    - Vehicle cycle simulation

2. **Performance**
    - Power curves
    - Governor curves
    - Lubrication oil consumption
    - Flow measurements
    - Heat balance
    - Emissions measurements

3. **Lubricants and Fuels**
    - Automotive lubricants
    - Marine lubricants
    - Black sludge formation
    - Intake valve deposits
    - Combustion chamber deposits

4. **Specialized Investigations and Testing**
   - Rig testing (e.g., bearings, antifreeze, erosion)
   - Simulated or environmental testing
   - Photoelastic stress measurements
   - Strain gauge testing
   - Flywheel burst testing

5. **Exhaust System Testing**
   - Vehicle cycle simulation
   - Steady state

6. **Catalyst Ageing**
   - Vehicle cycle simulation
   - Steady state
   - Accelerated ageing
   - Light-off efficiency tests
   - Sulfate release tests

Of course, there are other specific test types, but the preceding list covers the vast majority of test types that are likely to be encountered in the day-to-day testing undertaken by the research student, automotive engineer, and engine test technician.

Fully transient tests and, indeed, automatic mapping software programs are disciplines worthy of additional study. However, to glean the maximum useful repeatable data from all forms of transient testing, it is essential to have a full understanding of and experience with steady-state test types. Mathematical modeling of engine functions is an essential element in the design and development of new engine types. The accurate cross-correlation of modeled data with actual running data enables the leading manufacturers to move ahead of the opposition rapidly and to obtain clear market gains.

Understanding Durability Testing

This section is intended to provide an introduction to engine durability testing. A series of tests appropriate to a new engine development program is described. This is of particular importance for the test technician because it is extremely useful to have an understanding of the test cycle and where it fits into the overall development plan. Engine development is costly, and where prototype engine units are under test, they represent enormous investment and commitment from the manufacturer.

Before going into great detail, let us consider some basic principles. First, why is durability testing carried out? The objective of durability testing is to establish the probability of a given engine or product reaching its design life. Therefore, it is important that bench test conditions replicate and accelerate the in-service life conditions of the engine or product. Although a typical representative test cycle for all engine applications is not possible, four key failure areas would be assessed during testing. These are as follows:

1. **Mechanical failure**—This type of failure can be caused by the following:
   - Rubbing or sliding movement
   - Vibration induced by the firing strokes of the engine
   - Vibration induced by out-of-balance reciprocating or rotating masses

2. **Maximum heat input**—Under these conditions, maximum component operating temperatures are attained, and components whose durability is largely controlled by the thermal gradient are assessed. These components include the following:

   - Pistons (scuff-ring stick, etc.)
   - Valves and valve seats
   - Injector tip
   - Turbocharger

3. **Cyclic temperature variation**—These conditions occur when the engine is alternately operated between conditions of maximum and minimum heat input.

4. **Maximum mechanical and dynamic load**—With a turbocharged engine, this condition normally is encountered at maximum torque where cylinder pressure is at a maximum, and the lower operating speed reduces the extent of inertia relief. Components assessed include the following:

   - Small end
   - Big end
   - Mains
   - Pistons
   - Crankshaft

With dynamic load, this high inertia condition occurs at the maximum engine speed, normally governor run-out speed at no load. Maximum stresses are applied to the following:

- Valve and valve train
- Piston small end
- Main and big end bearings

Definitions

The popular terms "durability" and "reliability" often are confused. Therefore, before discussing durability tests in depth, it is necessary to define the correct meaning of each term.

- **Reliability**—The capability of an item to perform a prescribed task under defined conditions for a predetermined period of time

- **Durability**—The capacity of an item to reach a designed life

No clear-cut, definitive value for the term or span of an engine life exists, for with required maintenance, overhaul, and replacement of parts, this could be infinitely long. By common agreement, however, the automotive industry defines engine life as the time at which replacement of the key running components (i.e., crankshaft, bearings, pistons, and liners) is needed due to wear bringing them out of range of acceptable design limits. In the automotive industry, this life is further clarified as the time when the engine must be removed for overhaul.

Values for engine life are defined in terms of B10% and B50%, this being, respectively, the life at which 10% or 50% of the items under test have failed. If a high level of early failure rate is seen to occur, this is due to a combination of design shortcomings and the level of quality control during production. Note from the bathtub curve illustration (Figure 5.1) that the failure rate reduces to a point where it remains near constant (often

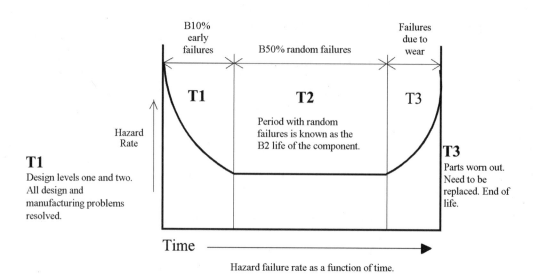

Figure 5.1 *The bathtub curve.*

referred to as the B2 life). During this constant rate period, failures normally are of a random nature and frequently are catastrophic, being the result of poor design, material specification, or manufacture, rather than being related to wear.

Reliability

The B10% life (the point at which 10% of the population has failed) normally will occur at some time during the constant failure rate area of the curve, although it can occur at any point in the bathtub curve. It is dependent on the level of random failures added to the number of early failures and hence defines the reliability of the engine.

Durability

Further increases in time show that the failure rate increases due to the natural wear process, with B50% life occurring during this period of increasing failure. Therefore, the B50% life determines the durability of the engine and is affected by the quality of the design, the specification of materials, and so forth.

The difference between durability and reliability can be demonstrated clearly by comparing the bathtub curves for different engines (Figure 5.2). Where two engines with the same B50% of 5.5 at 100,000 km exhibit different B10% lives, the engine with the higher B10% life would statistically be more reliable, although both statistically would have the same durability. For acceptability in the automotive market, the future rates of B10% and B50% life should not be less than 0.5–0.6.

The objective of durability testing is to establish the probability of a given product reaching its design life, which, in the case of an engine, may be specified as B50% = 7500 hours. This probability can be determined statistically only from a given number of engines and, to have a high level of confidence, can be determined only from historical field data from engines operating under similar conditions. This information is not available for new prototype engines; thus, some form of intelligent bench testing is required, which can be related to the likely service life of the engine.

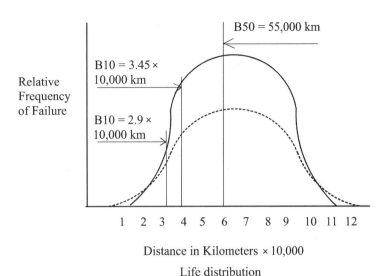

Figure 5.2 The difference between durability and reliability.

In-Cell Testing

To enable the results of bench tests to be related to expected service life, it is important that the operating conditions for the bench test be similar to those occurring in service. To cater for the majority of cases and to ensure that components are assessed under the most severe operating conditions, it is usual to specify a test cycle that incorporates one of the four extreme conditions already discussed in this chapter (i.e., mechanical failure, maximum heat input, cyclic temperature variation, and maximum mechanical and dynamic load). The engine build and installation on the test bench should replicate that of the engine installation in its vehicle. For a small number of engine applications (e.g., generating sets), it may be possible to predict with reasonable accuracy the probable operating cycle; hence, this requirement can be met. However, for most engines, the range of application is wide; consequently, the range of operating conditions and duty cycles is similarly wide. Therefore, a typical representative test cycle for all engines cannot be defined.

Analysis of the possible modes of engine operation reveals a number of discrete conditions that provide the most arduous operating conditions for various key engine components. Each of the conditions occurs in normal service, with the frequency and total time of occurrence depending on particular applications.

The specification of severe tests requires extreme care and normally can be achieved only by reference to historical data. It would be easy to specify tests of extreme severity that lead to early failure, but these cannot be related to normal service and are of dubious worth. Engine development on this basis results in the engine being over-designed and non-competitive. However, in the development of prototype engines, manufacturers will utilize tests at these known severe loading conditions to develop specific components. For example, in the development of a piston profile, it is common to operate the engine at continuous full load and speed (i.e., the maximum heat input condition). In parallel, other tests under alternative conditions on different engines would be carried out to develop the other critical components.

Early Development Phase Tests

In general, test times of 100–200 hours are found to be sufficient to screen new designs or to provide sufficient confidence to even consider longer-term (or extended) engine approval tests. Having completed the mechanical development of an engine using the previously described methods, manufacturers then need to undertake a specific durability test, normally referred to as a type approval test, which is related to the projected engine life. Such tests are vital to enable manufacturers to limit warranty claims and to maintain credibility and reputation within the marketplace.

Defining the In-Cell Test Procedure

No standards exist for type approval tests. Consequently, most manufacturers have developed their own methods that are based largely on their own historical service and past product data, as well as their knowledge of the applications and markets that they serve. In addition, some large users, who are able to define their expected duty cycle by drawing from their past experience, may specify their own approval tests. As discussed, manufacturers are unlikely to have a detailed knowledge of all the duty cycles to which their engines may be subjected in service. To legislate for the majority of cases and to ensure that components are assessed under the most severe operating conditions, it is usual to specify a test cycle that incorporates the extreme conditions previously discussed.

A possible cycle for durability tests is as follows:

- Rated load and speed—20 minutes
- Governor run-out speed, no load—10 minutes
- Maximum torque—20 minutes
- Idle—10 minutes

Increasing the Severity of the Test

Some manufacturers increase the severity of their tests by including test conditions that are more severe than those encountered in normal service. These may include overfueling, over-speed, advanced timing to increase cylinder pressure, extreme ambient conditions, and so forth. These tests are of value in reducing the time required to complete the test series, increasing the confidence level in the results when related to normal service, or a combination of both. However, the specification of such tests requires extreme care and normally can be achieved only by reference to historical data.

It is a simple matter to specify tests of extreme severity that lead to early failure and therefore cannot be related to normal service. Development on this basis results in the engine being grossly over-designed for its intended application and hence is non-competitive.

Early in a new engine program, at least two durability tests of 1000-hours duration should be undertaken. The objective of this work is to demonstrate that no major deficiencies are present in the design of the revised engine. Classical statistical analysis states that for tests carried out,

$$CL = 1 - R^n$$

where

CL = confidence level

R = reliability

N = number of tests completed without failure

From this, it can be seen that to demonstrate a reliability of 90% with a confidence level of 90%, a total of 22 tests is required (Figure 5.3). For the same reliability, one test gives a confidence level of 10%, while two tests increase this level to 19%. Therefore, it is clear that a statistical determination of reliability cannot be made from one or two tests with any reasonable degree of confidence.

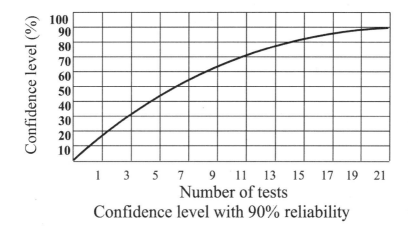

Figure 5.3 Confidence with 90% reliability.

When specifying the duration of durability tests, three aspects must be considered:

1. Verification that wear will not prevent the expected service life from being achieved

2. Verification that failure due to mechanical fatigue will not occur

3. Verification that failure due to thermal fatigue will not occur

The endurance limit for the typical engineering materials to be used in an engine family is variously quoted as between 10^6 and 10^7 cycles. For a four-stroke engine, a complete load cycle is considered to occur once for every two revolutions (every 720 crankshaft degrees). Taking the worst case of 10^7 cycles, this would occur after 166 hours of operation at a rated speed of 2000 rev/min.

Thus, a 1000-hour test based on mechanical fatigue considerations would be specified as follows:

- 2×10^1 cycles at full load and speed
- 4×10^6 cycles at idle
- 1.6×10^7 cycles at maximum torque
- 1.065×10^7 cycles at a governor run-out speed of 140
- A total of 5×10^7 cycles

This proposed 1000-hour test would demonstrate a high degree of confidence with respect to mechanical fatigue.

Thermal Stress

The maximum thermal stress on a given component occurs when it is operated over the maximum temperature range normally encountered. On an engine, this occurs when operating from full load to idle (or stop) and vice versa. The maximum induced stress in thermally loaded components is significantly higher than the maximum mechanically imposed stresses. Hence, the number of cycles to failure is lower than the endurance limit. In service, the number of maximum temperature-difference cycles that occur is several orders of magnitude less than the number of mechanical-load cycles. Hence, the number of thermal cycles required in a durability test to provide proof against thermal fatigue is reduced similarly.

Thermal Shock Testing

In the course of engine development, most manufacturers make use of thermal shock tests to assess component thermal fatigue and particularly cylinder head gasket durability. In these tests, the engine is cycled between full load and idle. During the idle mode, cold coolant is passed through the engine to rapidly reduce the component temperature (Figure 5.4).

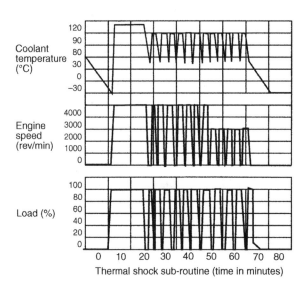

Figure 5.4 Thermal shock sub-routine.

The maximum component in-service thermal stress thus is applied in the minimum amount of time. (In fact, the cylinder head gasket is loaded beyond normal service conditions due to differential expansion between the cylinder head/crankcase and fixing bolts/studs.)

A possible test cycle consists of 5 to 10 complete thermal cycles per hour, at a total of 2000 cycles over the 100-to-200-hour period, and this accords with normal practice.

If necessary, more rapid cycles can be specified to reduce the total testing time, but care must be taken that temperature stabilization leading to metal relaxation, and hence full strain range, is achieved.

Combining Bench Testing with In-Field or Trials Testing

The ultimate durability assessment of an engine can be obtained only from service history. However, it is clear that this route cannot carry out development of engines alone due to both the time and the costs involved, hence the requirement for shorter-term bench durability tests. In formulating the time of these tests, there is a trade-off among cost, time, and confidence in the predicted service life.

As discussed, it is statistically preferable to increase the number of tests, such that the confidence level in the results of the bench test is high, rather than extend the duration of a limited number of tests. With the benefit of these results, manufacturers rely on a combination of historical service data and information from the concurrent field trials to establish the bench test correlation running tests.

Table 5.1 illustrates the extent of both the bench and field testing required to provide a statistically reliable estimate of durability for a number of engines.

TABLE 5.1
STATISTICALLY RELIABLE DIESEL ENGINE DURABILITY ESTIMATES

Description	Test Hours
Ford midrange diesel	
84 engines for bench test	130,000
121 field test vehicles	170,000
Hino EP100-11	
Various bench tests	30,000
John Deere 7.6-Liter	
Various engine bench tests	25,000
Field test vehicles	45,000
ISUZU 8.4-Liter	
Various engine bench tests	15,000
Field test vehicles	1,500,000

From this data, some manufacturers have been able to develop bench-service correlations that claim good accuracy; however, the commercial sensitivity of the data is such that full details rarely are published.

In general, with medium-duty diesel engines, these correlations often relate to total fuel used for the two types of testing being a ratio of 3:1 between bench testing and field testing. In addition, bench tests have been shown to be more severe than service operation; hence, a severity ratio also is required. This is quoted variously as between 5:1 and 10:1, depending on the test cycle.

Test Duration and Engine Life Comparison

An alternative is to consider the average life of the vehicle. Modern well-designed heavy-duty vehicles achieve 500,000 km before out-of-frame overhaul is required, and the trend is toward 1,000,000 km.

Average vehicle speeds for these heavy-duty vehicles worldwide usually are taken as 50 kph in mid- and Eastern Europe, and 80 kph in the United States. Therefore, taking the lower mileage figure, typical engine life ranges between 5000 and 8000 hours. In this chapter, we already established that these engines would have been subjected to bench durability tests of which 1000 hours duration is typical.

In assessing the results of a 1000-hour test, the overall criterion should be that the engine should appear to be in good condition and should continue to perform as specified.

Pass/Fail Checklist

At the end of the engine test, the following checklist should be applied to establish the outcome.

- No failure of the engine should occur during the test. This does not include minor failures such as fuel pipe breakage and so forth, which can be corrected in a short time and would not imply an out-of-frame repair service.

- Performance loss after the test period should be less than 5%.

- Oil consumption should show no significant increase and should remain within the target value. Blow-by should show no significant increase.

- There should be only minimal wear on the major running components (typically less than 0.025 mm on the cylinder liners and 0.01 mm on the crankshaft journals).

- Good component condition should exist, as judged by a skilled subjective assessment (i.e., freedom from cracks or other signs of distress).

- The specific brake fuel consumption should not increase by more than 5% from the initial value at any point through the running speed range.

Summary

What are the objectives of durability testing throughout the design–development phases of the new engine?

The objective of the work is to demonstrate the following:

1. No major deficiencies are present in the design of the revised engine, which will have a major impact on the service life and the reliability of the vehicle.

2. The tests are used to prove the quality and suitability of subcontractors' pre-production components.

3. The tests enable the automotive development engineer to verify that wear will not prevent the expected design service from being achieved.

4. Verify that failure due to mechanical fatigue will not occur.

5. Verify that emission and legislative performance will be maintained within the warranty life of the vehicle.

In all testing for both performance and durability, the following are some of the "golden rules":

- Suspect all results.
- Check and cross check calibration.
- Record all results, both the good and the bad.

Example of a Typical Test Procedure for a High-Speed Diesel Engine 250-Hour Validation Test

The following example is provided to illustrate typical test procedures. In this example, a high-speed diesel engine is undergoing a 250-hour validation test.

250-HOUR HIGH-SPEED DURABILITY TESTING INSTRUCTIONS

1.0 INTRODUCTION

1.1 The purpose of this test is to investigate the durability of a diesel engine and/or component parts at the maximum rated engine speed. The example given here is typical of an engine manufacturer's validation test. Note that when there are references to appendix numbers, these would be the manufacturer's data and are not included here.

2.0 GENERAL INSTRUCTIONS

2.1 The Requesting Engineer is to be advised immediately of any component failures and major leaks prior to rectification. Minor faults should be rectified at routine service periods. Before disturbing any components, note the relative positions of all affected components. Measure the residual torque of all affected hardware by the "crack-off and back to mark" method. If any "back to mark" torque drops below 30% of the original specification, the Requesting Engineer is to be informed.

2.2 Primary toothed belts must never be re-tensioned.

2.3 The timing belt cover must be fitted at all times during the test, except when access is required for diagnostic purposes.

2.4 After prolonged shutdown, warm the engine at 1500–2000 rpm half load for 5 minutes prior to resuming durability running.

2.5 The engine should be cleaned after any service or repair work, so that oil leakage points may be readily identified.

2.6 To fit in with facilities and manpower availability, the exact timing of checks and the test may be changed, provided that the Requesting Engineer has approved the changes. However, this does not apply to oil and filter changes.

2.7 All engine components and associated fastenings must be assembled using a calibrated torque.

2.8 Record all significant events and component failures on durability log sheets.

2.9 All service actions must be signed off by the relevant technician.

3.0 PREPARATION

3.1 Confirm engine shutdown fail-safes are on the following parameters. Also confirm their operation.

3.1.1 Speed

3.1.2 Torque/brake mean effective pressure (BMEP)

3.1.3 Coolant outlet temperature

3.1.4 Oil temperature

3.1.5 Exhaust temperature (pipe and pre-turbo)

3.1.6 Oil pressure

3.1.7 Boost pressure (if a turbocharger is fitted)

3.2 The cooling system is to be inlet pressurized as close as possible to the water pump inlet. The degas hose is to vent to the header tank above the coolant level. The hose should run level or upward for the whole length.

3.3 Confirm that the correct power correction factor as specified on the Engine Test Request is being used.

4.0 OPERATION

4.1 Conduct engine break-in to procedure as defined on the Engine Test Request.

4.2 If the engine has completed break-in on a different test bed, conduct a maximum performance wide open throttle correlation curve.

4.3 Perform any additional pass-off tests as specified on the Engine Test Request.

4.4 Configure the engine oil circuit to specification for durability testing, as specified on the Engine Test Request.

5.0 PRE-DURABILITY START OF TEST CHECKS

5.1 Check that an air cleaner is serviceable. Fit a replacement if required.

5.2 Check and record the valve clearances.

5.3 Examine the primary and auxiliary belts and record the average hot and cold Clavis gauge tensions.

5.4 Conduct a daily dyno calibration check.

5.5 Functionally check the injectors and record as-found nozzle opening pressure (NOP). Refit the injectors with new injector washers and seals.

5.6 Complete a maximum performance test.

5.7 Conduct a cylinder hot compression check, if not already recorded during the test procedure. Record the compression pressures, coolant outlet temperatures, and cranking speed. (Cranking speed should be a minimum of 180 rev/min.)

5.8 Run a governor run-out curve.

5.9 Run an oil pressure curve at speeds as applicable to the engine type if not already recorded during the test procedure. Maintain a bulk oil temperature of 110°C.

5.10 Record the crankcase blow-by if not already recorded during the test procedure, at idle, and at rated speed full load. Stop the engine and advise the Requesting Engineer if the levels are above the maximum listed in the Engine Test Request.

5.11 Fill the sump with new oil, and fit a new oil filter.

5.12 Flush out the cooler and associated pipe-work (if fitted).

5.13 Check for any leaks, and correct as necessary. Clean the engine and the test bed.

5.14 Ensure that all pre-durability tests, checks, and performance results are reviewed and agreed upon by the Requesting Engineer before commencing any durability test.

6.0 DURABILITY TEST

6.1 Record the total test bed hours and the durability running hours.

6.2 Set up the durability test conditions.

6.3 Start at zero durability hours.

6.4 Operate the engine continuously at the rated engine or speed specified on the Engine Test Request at full load for 250 hours.

6.5 Maintain the coolant outlet temperature as shown below or to the temperature specified on the Engine Test Request. Confirm with the Requesting Engineer.

Test Hours	Coolant Outlet Temperature
0-100	100°C ± 2°C
100-150	105°C ± 2°C
150-200	100°C ± 2°C
200-250	105°C ± 2°C

6.6 Maintain the coolant outlet fuel injection pump (FIP), fuel inlet, air filter inlet, and oil (sump) temperatures within operating conditions during the full load stages of the cycle as specified on the Engine Test Request.

6.7 Every 30 minutes, record all standard parameters.

7.0 SERVICE REQUIREMENTS

7.1 OIL CONSUMPTION TEST

7.1.1 During the first 3 hours in each 50 hours, conduct an oil consumption test at test speed/load condition. Record the results on the oil consumption data sheet.

7.2 EVERY 8 HOURS 20 MINUTES

7.2.1 Stop the engine. Position to No. 1 top dead center (TDC), and wait 5 minutes for the oil to drain down.

7.2.2 Dip the oil, and add oil as necessary to restore the level to the full dipstick mark.

7.2.3 Record the weight of all oil added.

7.2.4 Check/top up coolant as required. Record the volume added.

7.3 EVERY 25 HOURS/DAILY

7.3.1 Stop the engine. Position to No. 1 top dead center (TDC), and wait 5 minutes for the oil to drain down.

7.3.2 Dip the oil, and add oil as necessary to restore the level to the full dipstick mark.

7.3.3 Recheck the level after 5 minutes. Top up as required.

7.3.4 Record the weight of all oil added.

7.3.5 Check/top up the coolant as required. Record the volume added.

7.3.6 Review the engine performance using readings from full power stages. Stop the test and inform the Requesting Engineer if the power drops by more than 10% from the highest previous stabilized readings.

7.3.7 Wipe down the engine, and record the nature, position, and magnitude of all leaks.

7.3.8 Process the logged data. Report any deviation from the test specification to the Requesting Engineer.

7.3.9 Perform a daily dyno calibration check.

7.4 EVERY 50 HOURS

7.4.1 Stop the engine. Position to No. 1 top dead center (TDC), and wait 5 minutes for the oil to drain down.

7.4.2 Take 50 ml of lube oil sample from the sump, and label the sample jar with the dynamometer cell number, the engine number, the test hours, and the date. Record the weight of the oil sample (W1).

7.4.3 Dip the oil, and add oil as necessary to restore the level to the full dipstick mark.

7.4.4 Recheck the level after 5 minutes. Top up as required.

7.4.5 Record the weight of all oil added (W2).

7.4.6 Record the oil consumption, whereby $W_c = W2 - W1$.

7.4.7 Check/top up coolant as required. Record the volume added.

7.4.8 Review the engine performance using readings from full power stages. Stop the test and inform the engineer if the power drops by more than 10% from the highest previous stabilized readings.

7.4.9   Wipe down the engine, and record the nature, position, and magnitude of all leaks.

7.4.10 Visually inspect the crankshaft damper for signs of cracks in the rubber.

7.4.11 Conduct a blow-by test at the minimum load idle and the full load rated speed. Record the blow-by on the data sheet.

7.4.12 Check that the coolant flow is within limits, as specified on the Engine Test Request.

7.4.13 Drain the sump and fill it with a new test oil, as specified on the test request prior to draining. Top up as required to level, and record consumption.

7.4.14 Remove the oil filter, bag it, and label with the test detail. Fit a new filter.

7.4.15 Check/adjust the valve clearances. Record the results on the valve clearance data sheet.

7.4.16 Perform a daily dyno calibration check.

7.4.17 Process the logged data. Report any deviation from the test specification to the Requesting Engineer.

8.0   END OF TEST SCHEDULE AT 250 HOURS

8.1   Stop the engine. Position to No. 1 top dead center (TDC), and wait 5 minutes for the oil to drain down.

8.2   Take 50 ml of lube oil sample from the sump, and label the sample jar with the dynamometer cell number, the engine number, the test hours, and the date. Record the weight of the oil sample (W1).

8.3   Dip the oil, and add oil as necessary to restore the level to the full dipstick mark.

8.4   Recheck the level after 5 minutes. Top up as required.

8.5   Record the weight of all oil added (W2).

8.6   Record the oil consumption, whereby Wc = W2 – W1.

8.7   Check/top up coolant as required. Record the volume added.

8.8   Wipe down the engine, and record the nature, position, and magnitude of all leaks.

8.9   Visually inspect the crankshaft damper for signs of cracks in the rubber.

8.10 Examine the primary and auxiliary belts, and record the average hot and cold Clavis gauge tensions.

8.11 TAPPET CHECK

8.11.1 Check and record that the (cold) valve clearances are within tolerance.

8.11.2   Tappets must be reset before any of the final checks are completed.

8.12   FULL LOAD POWER CURVE

8.12.1   Perform a daily dyno calibration check.

8.12.2   Fit the instrumented injector to the cylinder No. 1 and the cylinder pressure transducer if applicable.

8.12.3   Run a full-load power curve at speeds as applicable for the engine type. Record the parameters.

8.12.4   Process the power curve data. If the performance is significantly out of specification (refer to the Test Request), inform the Requesting Engineer.

8.13   HOT COMPRESSION AND INJECTOR FUNCTION CHECKS

8.13.1   Perform a hot compression check. Record the compression pressures, the coolant outlet temperatures, and the cranking speed (cranking speed minimum 180 rev/min).

8.13.2   Functionally check the injectors, and record as-found nozzle opening pressure (NOP). Refit the injectors with new injector washers and seals.

8.13.3   Remove the instrumented injector and cylinder pressure transducer (if applicable) and refit the originals.

8.14   GOVERNOR HYSTERESIS TEST

8.14.1   Run a governor run-out curve (if applicable to the FIP type).

8.14.2   Process the governor hysteresis data.

8.15   CRANKSHAFT TORSIONAL VIBRATION TEST

8.15.1   Inspect the crankshaft torsional damper for signs of cracking in the rubber.

8.15.2   Conduct a torsional vibration test.

8.15.3   Process the data, and produce plots.

8.16   OIL PRESSURE CURVE

8.16.1   Run an oil pressure curve at speeds applicable to the engine type. Maintain a bulk oil temperature of 110°C.

8.16.2   Process the oil pressure curve data.

8.17   CRANKCASE BLOW-BY TEST

8.17.1   Record the crankcase blow-by at the following speeds. Stop the engine, and advise the Requesting Engineer if the levels are above the maximum listed in the Engine Test Request.

8.18   OIL CONSUMPTION TEST

8.18.1   Run a three-hour oil consumption test at the rated power.

8.19 NITROGEN PRESSURE TEST

8.19.1   Note and record the valve clearances. Do not reset.

8.19.2   Check and record the cold primary belt tensions.

8.19.3   Undertake a cylinder head gasket nitrogen pressure test.

8.19.4   A nitrogen pressure test must be conducted after the completion of all engine running.

8.20 FINAL CHECKS

8.20.1   Take a one-liter oil sample.

8.20.2   Remove the oil filter, bag it, and label with the test detail. Fit a new filter.

8.20.3   Remove the air filter, bag it, and label with the test detail.

8.20.4   Check the fuel filter for leakage. Remove, bag, and label it with the test details (if the vehicle filter has been fitted for the duration of the test).

8.20.5   Remove the power take-off unit, and perform a rear crankshaft oil seal leak examination.

8.20.6   Examine the whole engine, including under the timing belt cover, for leaks and any signs of failure. Record the findings.

8.20.7   Remove any special test bed fittings. Replace the original parts low-pressure fuel pipes and hardware, vehicle thermostat, and coolant sender (if fitted).

8.20.8   Process the Engine Build and Test Record sheets and file them with the Engine Log Sheet and daily dyno calibration in the Supplementary Log. Do not dismount the engine until all end-of-test data have been reviewed by the Requesting Engineer.

8.20.9   Remove the engine from the test bed. Ensure that all components removed prior to the start of testing either are refitted to the engine or are labeled and returned with the engine.

8.20.10   Ensure that test oil samples and filters are returned with the engine.

8.20.11   Ensure that all failed components have a concern note raised against them.

Note that the order of completion of the final checks is discretionary. However, the valve clearance must be checked/adjusted before any of the performance tasks are completed, and the nitrogen pressure test must be conducted after all engine running is complete.

GENERAL NOTE

Prior to removing the engine from the test bed, please affix the labels.

Highlight any gas or liquid leaks noted, and any possible problem areas noted in the course of this test.

Master Service Record Sheet

Master service record sheets cover the operator's instructions throughout the test and ensure total continuity over the life of the test. Examples of such sheets are given below.

SERVICE ACTION	0.00 Hrs	8.20 Hrs Min	16.40 Hrs Min	25.00 Hrs Min	33.20 Hrs Min	41.40 Hrs Min	50.00 Hrs Min	58.20 Hrs Min
Conduct break-in as detailed on Engine Test Request (ETR)								
Carry out additional service tasks								
Start oil consumption test								
Stop oil consumption								
Test and calculate result								
Take 2 off 50-ml oil samples								
Check oil level and record additions								
Check coolant level and record additions								
Carry out daily dyno								
Calibration check								
Process logged data and review performance								
Inspect engine for leaks								
Inspect torsional vibration (TV) damper								
Coolant flow check (1)								
Record blow-by at idle and at rated speed full load (1)								
Drain engine oil								
Refill with new test oil								
Remove bag and label oil filter. Fit replacement (2)								
Flush out oil cooler								
Check for leaks								
Remove bag and label fuel filter. Fit replacement (2)								
Check/adjust coolant specific gravity								
Carry out crankshaft TV survey								
Record hot and cold belt tensions								
Check/adjust valve clearances								
Remove bag and label air filter. Fit replacement (2)								
Record full load power curve (1)								
Record oil pressure curve (1)								
Check/reset MNLS (1)								
Record governor hysteresis curve (1)								
Record hot cylinder compressions (1)								
Take 2 off 1-liter oil samples								
Carry out nitrogen pressure test								
Carry out rear seal examination								

(1) These pre-test tasks need not be repeated if already completed during engine break-in and pass-off (BIPO).
(2) New air, oil, and fuel (if vehicle part is used) filters must be fitted before the test starts.

Interpretation of the Test Results

How can one determine if the completed test is good or bad? After the final performance test of a validation series, the engine should be completely stripped, examined, and the conditions of the various parts recorded. The main wearing parts should be photographed. Typically, wear rates and performance losses shall be within the following specified limits:

Engine Component	Light-Duty Diesel	Heavy- and Medium-Duty Diesel
Cylinder bore	0.025 mm	0.05 mm
Top piston ring gap increase	0.20 mm	0.3 mm
Piston side clearance increase	0.10 mm	0.15 mm
Piston skirt clearance measured at grading point	0.05 mm	0.10 mm
Valve seat depth increase	0.10 mm	0.30 mm
Tappet wear	0.05 mm	0.10 mm
Cam lift	0.05 mm	0.15 mm
Valve clearance change	0.20 mm	0.30 mm
Chain elongation	0.5%	0.5%
Crank journals	0.025 mm	0.025 mm
Crank pins	0.025 mm	0.025 mm

No significant oil leakage is allowed through the sump gasket, valve gear covers, front and rear crankshaft seals, and the cylinder head gasket. Also, the cylinder head gasket should be in good condition.

Total Preventative Maintenance: Daily Test Bed/Cell Checks

It is the responsibility of all test technicians to ensure that the test beds they are running are in a clean and tidy condition at all times, and that all test-cell-related equipment is in good operating order. Where instrumentation equipment is required to have calibration tractability, then the test technician must ensure that the calibration due date has not passed. Within the test cell, the main items to be checked on a daily basis are as follows:

- Check the bedplate holding down fasteners and the condition of the anti-vibration pads.

- Check the dynamometer mounting plate holding down fasteners and the condition of the anti-vibration pads.

- Check all dynamometer-to-mounting-plate fasteners.

- Check all engine-pallet-to-bed-plate fasteners.

- Check all engine pallet fasteners (e.g., cross members, legs, supports, brackets).

- Check the engine mounting rubbers for good condition. (Replace any that are suspect, e.g., oil contaminated, cracked, split). Should a replacement be fitted, then the alignment, pitch, and roll should be rechecked.

- The dynamometer water filters should be cleaned at least once per shift during a scheduled engine stop period.

- All thermocouples should be checked for good condition and connection.

- All LEGRIS piping should be checked to ensure good condition, and any trap pots in the circuit should be drained. (Some tests require the trap pot contents to be weighed before discarding the contents.)

- Pressure lines (i.e., oil, fuel) should be checked for good condition.

- Exhaust systems should be checked for cracks, splits, and so forth.

- Check the throttle actuation for zero to full span operation.

- Prop shafts, where in use, should be lubricated at regular intervals (once per 24 hours).

- Spot cooler fans should be checked for good condition. Any found to be faulty should be repaired or replaced.

- Once all in-cell work has been completed, the engine and bedplate should be cleaned.

- Finally, when the engine is in a state of readiness to return to the test, then the test cell floor should be cleaned.

Recordkeeping

Reasons for Keeping Records

Probably one of the most important parts of the test technician's role is to keep and maintain complete and accurate records of engine information.

Whether the engine has run through the test program without any faults occurring, or the test has not been completed due to engine failure, the information must be recorded. It is on this information that the customer bases his or her engine/product viability as a marketable commodity. The customer places confidence in the ability of the test technician to gather all the relevant information on his or her behalf; therefore, the test technician must record every incident that occurs regarding a test engine, no matter how small or trivial that event may seem.

Types of Records

The main types of records to be kept span the following areas.

Sales and Marketing

On initial contact with the customer, the sales and marketing team and the customer decide on the best test program through which to run the engine, so that the customer receives the most relevant information feedback from testing of the product.

A customer requirements document should be completed. This becomes the first traceable documentation for that customer.

Engineering

Once the decision has been made about which is the best test program to use, the Engineering Department allocates the engine test program to an engineer, who then is responsible for setting up that program to the customer's specifications. The engineer will open a file that will contain all relevant information and the history of the engine as it passes through the test house. This will include records such as engine break-in records, performance records, service records, engine failure records, customer test specification changes, test bed calibration records, and so forth.

Bed/Pallet Build

Once the engine is delivered to the test house, the bed/pallet build team is issued a copy of the customer requirements documentation.

From the information supplied in this document, the engine is mounted on a suitable bed/pallet, in accordance with the customer's specifications (e.g., engine pitch and roll angles, instrumentation requirements, dress state).

When the engine and pallet assembly have been completed, it is then transported to the engine test cell and installed into its test bedplate.

Installation

The installation of the engine is carried out either by the test technician or the bed/pallet build team.

In the case of a new customer's engine being installed for its first test, the instrumentation technician will set up the instrumentation.

The instrumentation technician also may be required by the customer to perform a full pre-test calibration check of all engine monitoring equipment as part of the test program.

Customer requirements may include a completed installation check sheet, calibration record sheets, and so forth.

Running

The Requesting Engineer will produce all the record sheets required throughout the test program for running the engine. These include the following:

- Break-in procedure
- Service schedule (all engine adjustments are set to datum)
- Performance curves (power, oil pressure, coolant flows, blow-by, ignition)
- Durability/high-speed tests (or both), with specified services and power checks
- End-of-test service schedule (again all engine parameters are set to datum)
- Final performance curves

Prior to removal of the engine from the test bed, the Requesting Engineer should check the final performance figures and verify that they are acceptable. It may be that a re-run of one, or all, performance curves is required. All of the preceding information is to be recorded either on handwritten sheets or computer data acquisition. The data acquisition material then can be reproduced by the Requesting Engineer to form part of the customer's final test report.

Reporting

During the course of the engine test program, the Requesting Engineer may be in contact with the customer on a day-to-day basis. In this way, the customer will be kept up to date with the progress of work on his or her engine. In the event of continual engine component problems, the Requesting Engineer should inform the customer of this fact, along with any corrective action the Requesting Engineer may feel would be of assistance to the customer. The information from the test program is prepared in full engine test report format by the Requesting Engineer. Should the Requesting Engineer have any valid recommendations, then these should be included in the test report so that the customer can evaluate these.

Sales and Marketing

The sales and marketing team will have the final contact with the customer, usually to gain further work from the customer but also to conduct a follow-up survey. This is to establish whether the customer is satisfied with the service he or she has received and to record any recommendations on how to improve the service that was provided.

Fault Diagnosis

Fault diagnosis is an area where valuable test time can be lost at an alarming rate. It also is an area where the technician's understanding of engine technology is best shown or found to be lacking. Requesting Engineers should note that it is imperative that the test technician be given training on new technology as it is introduced into the industry. Good, relevant, and effective training will ensure that the test house has knowledgeable and capable test technicians. The test technician must be able to diagnose an engine, test bed, and test-cell fault as quickly as possible. This keeps costs down for the customer and the test house. The diagnosis of faults is best achieved through a process of elimination and must be done in a logical step-by-step procedure. An example of this procedure is given below.

In the following example, an engine will not start but will crank after it has been fitted to a test bed. However, do not immediately condemn the ECU. First, check the more logical items that make an engine run, as follows:

- Does it have fuel?
- Is the fuel reaching the engine?
- Is there an ignition spark?

There may not be a fault with the engine itself, or it may be something simpler.

Has the induction manifold transportation plug been removed?

Have any of the test-cell emergency stop buttons been operated?

The test technician's knowledge of the test bed and cell must be of a reasonable level. The technicians must make themselves aware of the test cell environment, in much the same way as the vehicle mechanic must understand vehicle fuel and electric supplies to the engine. Test technicians should *never assume anything*.

Reporting

This part of the technician's job is most often neglected and can, if performed with care and diligence, be of the most benefit to all concerned. The customer requires the information and continuous feedback from the test cell, so that he or she may see any inherent problem areas and, if necessary, may make modifications before the engine is tested further or is released into production. The record of types and frequency of faults that occur with any engine can be used by the test technician to shortcut the fault-finding process.

For example, an engine running at 5000-rpm full load begins to misfire. The technician shuts down the engine to investigate the problem. It is found that battery voltage was down to 9.5 volts. Testing the battery with a voltage-drop tester shows that the battery cells are OK.

After changing or recharging the battery, further investigation reveals that the alternator terminals are separated from the wiring. After replacing the terminal ends, the engine is restarted; a check on the charging of the battery now shows it to be correct. The engine goes back on test.

Over the next few days while other shift personnel run the engine, the same report is filed another three times. The shift personnel are able to check the engine faults records and see at a glance that the symptoms were the same and what corrective action was taken.

The Requesting Engineer in charge of the project, by checking the fault records, then can take certain steps to ensure that the fault does not lie within the engine-to-test-bed setup (e.g., engine-to-dynamometer alignment causing excessive vibration) or to ensure that all of the manufacturer's cable holding clips are fitted in the correct places and do not stretch the cable routing.

Once the Requesting Engineer is satisfied that the fault does not lie within the test house procedures and the fault definitely is with the customer's product, he must inform the customer at the earliest opportunity. The fault record can be in the form of a test bed running log or diary that remains with that specific test bed for the duration of the engine test project.

Chapter 6

Spark Plugs

Let us for a moment consider the working environment of spark plugs. They are designed to cope with the following conditions:

- Reliable operation for at least 30 million cycles

- Normal operating temperature between 350 and 900°C, with possible temperature delta between these extremes in less than 4 seconds

- Between 14 and 25 kV electrical stress

- Running at 6000 rev/min, that is, every 20 mS almost instantaneous pressure from negative to 1000 psi (6.9×10^6 Pa)

- A corrosive environment, including sulfur, bromine, phosphorus, and, to some degree, tetraethyl lead

- The need for abrasion resistance against high-velocity particles

The heat generated by the combustion of gas and air in a modern engine can result in temperatures of more than 2500°C inside the combustion chamber. Outside the engine, at the terminal end of the plug, the air temperature can be sub-zero in winter. A temperature of 2500°C would melt all the metal around it if the flame heat were continuous. However, the flame lasts for only one stroke of the engine. At idle, 600 rev/min, it lasts for only 0.05 seconds; at 6000 rev/min, it lasts for 0.005 seconds. In the four-stroke engine, there are three strokes without flame (i.e., induction, compression, and exhaust) for each power stroke, during which the heat generated in the power stroke is conducted to the cooling system via the engine walls.

The new generation of 42-volt ignition systems has given rise to high-kilovolt discharge plugs. New detergent additives have been added to fuels, and various methane- and hydrogen-based gaseous fuels are being utilized, which require higher-kilovolt spark discharge, and spark plugs are being designed to last for the life of the power unit. This is a truly daunting task for spark plug designers and development engineers.

Much of spark plug construction is shown clearly in the Figure 6.1, but some aspects are worthy of comment. A spark plug serves the function of conducting high-tension current into the combustion chamber to provide ignition of the fuel/air mixture. Sealing is a major problem, and this is undertaken by the composite metal/glass conducting seal inside the insulator, as well as by the metal gasket between the insulator and the mild steel shell. These seals play an important role in the performance of spark plugs, in addition

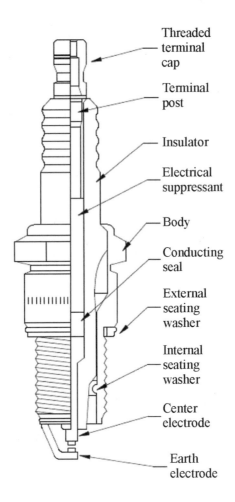

Figure 6.1 Spark plug construction.

to sealing gas pressures. They provide the main paths by which the heat may transfer into the bulk of the cylinder head from the spark plug components (Figure 6.2).

The heat generated at the firing end (tip) of the spark plug when the fuel/air mixture is ignited is dissipated in the engine by conduction via the ceramic insulator nose, the central electrode, and the mild steel shell.

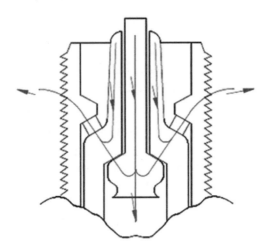

Figure 6.2 Heat conduction paths, cold plug heat flow.

The metal components of the construction transfer heat more rapidly than the ceramic insulator, with the relevant thermal conductivities differing by approximately a factor of two. If the heat is not removed efficiently, with subsequent firings the ceramic tip (i.e., the nose) of the spark plug soon attains a temperature at which it is able to pre-ignite the fuel/air mixture before full compression is achieved, and prior to the timed electrical discharge taking place. This pre-ignition results in a drastic loss of power and increased levels of gaseous emissions. If allowed to continue, the increasing retention of heat can lead to permanent damage of the spark plug, piston, cylinder head, and catalytic converter formulation.

This ability of the spark plug to remove heat from its firing end determines its suitability for a particular type of engine (i.e., the spark plug heat range). It is evident that a standard quantitative measurement of this ability is desirable to relate one plug to another in the range of designs in existence and to correlate the products of different manufacturers. The current method used exclusively in the industry is known as the pre-ignition rating of spark plugs.

Spark Plug Ratings

The method used for ascertaining a spark plug rating is to use a single-cylinder four-stroke engine, whose essential nature is constant speed running, with changes in power brought about by supercharging. Supercharging occurs when air or the air/fuel mixture is presented to the cylinder at a pressure that is higher than the atmospheric pressure. Because of this higher pressure, the air supplied to the cylinder has a higher density and can absorb more fuel vapor. This increases the power output.

The performance and rating of spark plugs can be defined as follows and is demonstrated in Eqs. 6.1 through 6.4.

Indicated work (W_i) is the net work produced in the cylinder by the gas acting against the piston during the compression and expansion strokes. Therefore,

$$W_i = W_b + W_f$$

where

W_b = brake work

W_f = friction work

Brake work is the work available at the engine shaft, whereas friction work includes mechanical friction and, in four-stroke engines, pumping work during the exhaust and intake strokes.

Thus,

$$W_i = n \int p \cdot dV \tag{6.1}$$

(for compression and expansion strokes only), where

> n = number of cylinders
>
> p = net pressure acting on the piston
>
> V = cylinder volume

For a four-stroke cycle,

$$Wb = 4\pi T \qquad (6.2)$$

where T is the average torque at the engine output shaft and

$$Wf = Wi - Wb \qquad (6.3)$$

which is the indicated mean effective pressure (IMEP). The indicated work produced per cycle per unit of engine displacement equals the constant pressure that would produce the same work if it acted on the piston during only the power stroke.

$$\text{IMEP} = \frac{2KW}{D} = \frac{4\pi KT}{D} \qquad (6.4)$$

where

> W = work per cycle
>
> T = torque
>
> D = engine displacement
>
> K = conversion factor

With this data, one can define the performance of a spark plug. The spark plug rating is that IMEP value obtained on the rating engine at a point when the supercharge pressure is 34 mb below the pre-ignition point.

Running a Spark Plug Rating Test

For a given spark plug installed in a rating engine, by gradually increasing the amount of supercharging and adjusting the fuel mixture strength to give optimum temperature at each setting, the plug experiences higher and higher temperatures until it begins to run into pre-ignition (indicated by a rapid rise in the measured plug temperature). As pre-ignition occurs, the fuel supply instantly is cut off, preventing uncontrolled temperature rise and possible damage to the engine. When stable operation is obtained (34 mb of supercharge boost below the pre-ignition point for three minutes), the torque is measured, allowing an IMEP value to be calculated according to Eq. 6.4. At any fixed set of engine conditions, there is a definite boost–IMEP relationship, which is a straight-line function (Figure 6.3).

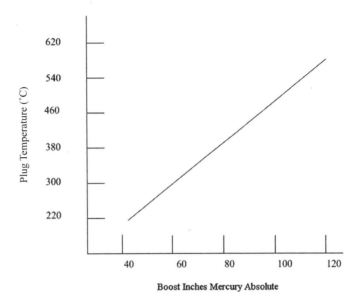

Figure 6.3 Engine boost against output for optimum plug temperature at 30° spark advance.

Various IMEP values are required because of the different demands on engine performance. For example, a racing car must run at high temperatures for maximum efficiency and power output over a relatively long period. The best spark plug for this environment would be one that could dissipate heat rapidly to the engine mass. For a hot engine, a plug is needed that remains as cold as is necessary to prevent pre-ignition. Therefore, a cold plug is required (Figure 6.2).

The IMEP value of a cold plug would be relatively high. However, if the same plug were used on a family sedan that is used on short urban journeys and thus never fully warms up, combustion deposits soon would build up and lead to misfiring. In this situation, a plug with a relatively low IMEP value is required that does not dissipate the heat as readily and that has an operating temperature that will be sufficiently high to burn off the combustion deposits. Thus, a cold engine calls for a hot plug (Figure 6.4). Other terms that are used are "hard," which is the equivalent of cold, and "soft," which is the equivalent of hot.

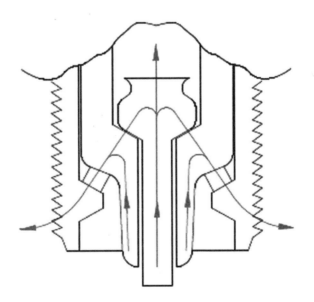

Figure 6.4 Hot plug heat flow.

However, the much shorter heat path of the cold plug can cause problems in that there is much less area of the ceramic insulator exposed to the cylinder. Under certain engine conditions, such as cold start with full enrichment, carbon combustion deposits build up on the nose of the spark plug, offering a leakage path to earth for the current. This leads rapidly to a situation wherein the spark plug ceases to function, known as cold fouling. Substitution by a normal spark plug of lower IMEP rating, thus providing a longer nose to overcome this problem, may be unsatisfactory due to the reduction in maximum safe operating temperatures. However, the spark plug designer and the engine development engineer have one or two options available to satisfy the provision of a greater surface area of the insulator nose while maintaining the IMEP rating.

Recent years have seen the adoption of center electrodes containing a core of copper. These electrodes are more thermally conductive than the typical nickel-alloy types and enable a longer electrode and ceramic nose to be employed. This solution elegantly achieves the aims discussed previously, that is, larger surface area for the same heat rating. An alternative means of increasing the heat removal rate is to eliminate the air gap between the ceramic nose and the center electrode by filling the space with a refractory cement material (Figure 6.5). This normally is achieved by the application of a vacuum to the tip of the spark plug, thus allowing the re-establishment of atmospheric pressure to force the cement into the evacuated space. Air can be an effective thermal insulator, and its replacement by the solid cement enables the heat to transfer from the ceramic insulator much more efficiently by conduction into the center electrode along its entire length and hence be more readily dissipated.

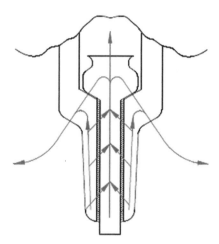

Figure 6.5 Improvement in heat transfer rates by vacuum cementing.

Because most of the heat conduction takes place by way of the electrodes, the use of materials that are more efficient in this respect obviously must be considered. However, the employment of, say, nickel in place of nickel alloy increases heat dissipation, and hence IMEP, but at the expense of electrode durability. This is due to more rapid chemical, electrical, and mechanical erosion of the softer material, despite its higher temperature tolerance.

One problem of which the designer must be aware in producing a very cold plug is a possible change in the site from which pre-ignition may occur. Normally, initiation of pre-ignition will take place from the overheated ceramic nose, which is unable to dissipate the heat rapidly enough, provided there are no incandescent sharp burrs or a glowing protrusion of combustion deposits present to pre-empt such occurrence. As the IMEP value is increased, the removal of heat from the nose becomes more efficient, lessens the chance for overheating of the ceramic body to occur, and shifts the emphasis to the side electrode, which then becomes the prime site for pre-ignition. The plug then is said to rate off the side wire instead of the nose. Attempts to control this phenomenon center on improving the heat removal rate of the side electrode either by a change of material, as already discussed, or by cutting back the side wire to provide a shorter path.

The material changes within the ceramic insulator itself can have an effect. The almost universal use of aluminum in this application is the result of historical changes that are admirably chronicled. (See J.V.B. Robinson, *History of the Sparking Plug. Motor Management,* 1971, Vols. 6 and 7, and J.S. Owens *et al., Development of the Ceramic Insulator for Sparking Plugs,* American Ceramic Society, 1977, Vol. 56.) The various manufacturers employ compositions ranging from approximately 88 to 95% aluminum. Changes in aluminum content toward the lower end of this range are attractive to manufacturers for several reasons. The major benefit is the large savings in energy usage arising from the lower sintering temperature, which is brought about by the replacement of aluminum with less refractory additives. Additional savings come from increased kiln and kiln furniture life at the lower firing temperatures, as well as reduced material costs. However, the design engineer then must accommodate altered material properties, the most important of which, from his point of view, is the reduced thermal conductivity—because the alumina content decreases in the annular cross-sectional area of the ceramic nose to retain the heat transfer ability. Such thickening of the nose has the effect of aiding anti-fouling characteristics of the design by increasing the surface area. Two or more of these changes can be combined by the spark plug development engineer to achieve a degree of fine-tuning of the rating values.

Changes in spark plugs have been caused by the demands for longer life and improved ignitability in today's lean-burning modern engines. It is critical that each cylinder fires to prevent unburned fuel from reaching the catalytic converter and causing damage. Variations in burn rates on a cycle-to-cycle basis are critical when new emission regulations are reviewed; a few parts in one million of trace emission gases can be the difference between a regulatory pass or fail.

With engine manufacturers striving toward lean-burn engines to improve fuel consumption, the possibility of misfire becomes a problem. In addition, manufacturers are striving to produce an engine in which the spark plugs will never need to be changed. A high voltage with large gaps appears to be the answer.

It has been proven that a thin electrode will improve ignitability; however, a thin electrode made from nickel would erode quickly and fail. Therefore, spark plug manufacturers have necessarily turned to precious metals to withstand the harsh environment.

The introduction of precious metals such as platinum (Figure 6.6) and iridium give spark plugs the added benefit of long life. Note that spark plugs historically have been regarded as consumable, with a limited service life, and as such required to be changed at regular intervals. Some NGK spark plugs were designed with a patented V-groove for improving ignitability in nickel alloy electrodes. This style of spark plug incorporates a 90° V-groove in the center electrode and ensures that sparking occurs at the periphery of the electrode, thus enhancing ignitability. The introduction of 42-volt ignition systems means that even higher spark plug voltages can be applied, leading to larger plug gaps (up to 3 mm) that have extended life.

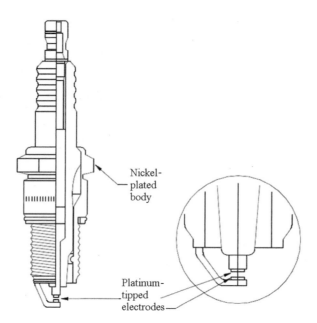

Figure 6.6 *Platinum-tipped electrodes.*

As stated, the use of precious metals in spark plugs has increased their service life. The use of iridium spark plugs is only starting with Japanese car manufacturers, who are finding them ideal for very low emission engines.

Radio Frequency Interference

All modern engines require the use of resistor-type spark plugs. A resistor-type spark plug is one that incorporates a 5-K ohm resistor to suppress the ignition noise generated during sparking. Radio frequency interference (RFI) is commonly exhibited by the crackling sound coming from the car radio, and there are now international standards covering RFI, which is considered a type of pollution.

Because RFI also can cause premature failure to other electronic components in a modern vehicle (e.g., the ECU), it is important that resistor spark plugs are used to prevent this possibility.

Summary

We have discussed the selection of heat range and its importance, and why this is vital to ensure optimum performance of spark plugs. The optimum operating temperature of a spark plug is between 450 and 870°C. Spark plug tip temperatures outside this range can occur when an incorrect heat rating is selected.

When the heat rating is too high, the spark plug temperature remains too low and causes deposits to build up on the firing end. These deposits offer an electrical leakage path that gives rise to a loss of sparks.

When the heat rating is too low, the spark plug temperature rises too high and induces abnormal combustion (pre-ignition). This leads to melting of the spark plug electrodes, as well as piston seizure and erosion. Many plug manufacturers have pioneered the use of a copper-cored electrode, with NGK spark plugs being the first to do this in 1958. This enables a spark plug to heat quickly and to dissipate heat quickly, giving an ultra-wide heat range.

Because spark plugs are positioned in the cylinder head of an engine, their analysis can give a good indication of how an engine is operating. Therefore, it is important to examine any spark plugs that have been removed from engines in order to learn from the visual evidence. Oxidization, metal deposits on the ceramic, and copper migration from gaskets all tell a story.

When such data are reviewed alongside the rate of cylinder pressure rise against the crankshaft angle (combustion analysis) and a measurement of the gaseous and particulate emissions, then the development engineer has a very good tool.

To date, spark plugs have been used only in gasoline or gas-based fuels. However, new emissions regulations may force diesel engines to control the point of burn initiation by utilizing spark plugs.

Chapter 7

Exhaust Gas Emissions and Analysis

Exhaust Gas Emissions

Many countries have legislation aimed at improving air quality by reducing the emission of harmful gases from motor vehicles. These harmful gases can be categorized into four groups:

Group A—Gases that can cause death or injury within minutes

Group B—Gases that can cause death or serious illness with prolonged exposure

Group C—Gases that can create minor health problems or are a nuisance

Group D—Gases associated with global warming

Group A—Gases That Can Cause Death or Injury Within Minutes

Carbon monoxide (CO) is a chemical compound of carbon and oxygen with the formula CO. It is a colorless and odorless gas that is approximately 3% lighter than air and is poisonous to all warm-blooded animals and many other forms of life. When inhaled, it combines with hemoglobin in the blood, preventing absorption of oxygen and resulting in asphyxiation. Carbon monoxide normally is formed by an incomplete chemical reaction and is especially probable when that reaction forms quickly, as in a car engine. For this reason, car exhaust gases contain harmful quantities of CO, sometimes several percent, although anti-pollution devices are intended to keep the level below 1%. As little as one-thousandth of 1% of CO in air may produce symptoms of poisoning, and as little as 0.25% may prove fatal in less than 30 minutes. Carbon monoxide is a major ingredient of air pollution in urban areas.

Because it is odorless, CO is an insidious poison. It produces only mild symptoms of headache, nausea, or fatigue, followed by unconsciousness. A car engine running in a closed garage can make the air noxious within a few minutes.

Groups B and C—Gases That Can Cause Death or Serious Illness with Prolonged Exposure and Can Create Minor Health Problems or Are a Nuisance

Smog

Smog is a mixture of solid and liquid fog and smoke particles. It forms when humidity is high and the air is so calm that smoke and fumes accumulate near their source. Smog reduces natural visibility and often irritates the eyes and respiratory tract. In dense urban areas, the death rate may rise considerably during prolonged periods of smog, particularly when a process of heat inversion creates a smog-trapping ceiling over a city. Smog occurs most often in and near coastal cities and is an especially severe air pollution problem in Athens, Los Angeles, and Tokyo.

Internal combustion engines are regarded as one of the main contributors to the smog problem, emitting large amounts of contaminants such as unburned hydrocarbons and nitrogen oxides. However, the number of undesirable components in smog is considerable, and the proportions are highly variable. They include ozone, sulfur dioxide, hydrogen cyanide, and hydrocarbons and their products formed by partial oxidation.

So-called photochemical smog, which irritates sensitive membranes in mammals and damages cell structure in plants, is formed when nitrogen oxides in the atmosphere undergo reactions with the hydrocarbons energized by ultraviolet and other types of radiation from the sun.

Nitrogen Oxides (NOx)

Nitrogen oxides are produced by the reaction of nitrogen and oxygen in the air at very high temperatures within combustion engines. They exhibit the following traits:

- They dissolve readily in water to give nitric and nitrous acids, causing acid rain.
- They are linked to the depletion of stratospheric ozone.
- They are greenhouse gases.
- They cause photochemical smog.

Sulfur Dioxide (SO_2)

Sulfur dioxide is a colorless gas that forms when sulfur burns in air. In an automobile engine, it is produced by sulfur impurities in the fuel. Sulfur dioxide dissolves in water to give a mixture of sulfuric and sulfurous acids, thus increasing the acidity of rainwater. It also is linked to some lung conditions.

Unburned Hydrocarbons and Particulates

Benzene, formaldehyde, carbon particles, 1,3-butadiene, and other organic chemicals are produced by combustion engines. These substances pose a health risk, and some products, such as carbon particles, are linked to asthma and other lung conditions. In particular, benzene is carcinogenic.

Cancer-Causing Agents

Particulates and other airborne hydrocarbons may promote cancer under certain conditions. The particles emitted by motor vehicles are small enough to penetrate deep into the lungs and therefore may lead to medical conditions similar to those caused by smoking.

Group D—Gases Associated with Global Warming

It has been known since 1896 that carbon dioxide (CO_2) helps stop the sun's infrared radiation from escaping into space and thus functions to maintain the earth's relatively warm temperature. This is known as the greenhouse effect. The question is whether the measurably increasing levels of carbon dioxide in the atmosphere over the last century will lead to elevated global temperatures, which could result in coastal flooding (through a rise in sea level) and major climatic changes. Such conditions would have serious implications for agricultural productivity. Since 1850, there has been a mean rise in global temperature of approximately 1°C (1.8°F), but this rise could be part of a natural fluctuation. Such fluctuations have been recorded for tens of thousands of years and operate in short-term and long-term cycles. The difficulty of distinguishing human-made causes of carbon dioxide emissions from natural sources is one reason why governmental legislation regarding control of these emissions has been slow in coming. However, the potential consequences of global warming are so great that many of the world's top scientists have urged immediate action and have called for international cooperation on the problem.

Carbon dioxide is a greenhouse gas that contributes to the greenhouse effect by raising the temperature of the earth. This is known as global warming. It also lowers the pH of rainwater because it will dissolve in water to give carbonic acid. Levels of carbon dioxide in the atmosphere have increased by 12% in the past 100 years, mainly due to extensive burning of fossil fuels and the destruction of large areas of tropical rainforest.

Simple Combustion Theory, Ideal Combustion, and Stoichiometry

Of What Is Fuel Composed?

Petroleum fuels consist largely of hydrogen and carbon. Hydrogen is a light atom with a relative mass of 1; carbon is a heavier atom with a relative mass of 12. The atoms of hydrogen and carbon are chemically bonded together to form hydrocarbon molecules. The simplest hydrocarbon molecule is methane (CH_4) (Figure 7.1).

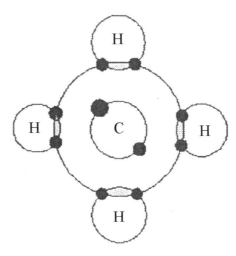

Figure 7.1 Methane.

One molecule of methane consists of one carbon atom bonded to four hydrogen atoms. The relative mass of a carbon molecule is 12, and the relative mass of a hydrogen molecule is 1; therefore, the relative mass of a methane molecule is $12 + (4 \times 1) = 16$.

Hydrogen contributes 4/16 or 25% to the mass; therefore, the hydrogen mass fraction is 25%. Carbon contributes 12/16 or 75% to the mass; therefore, the carbon mass fraction is 75%.

Methane is a relatively simple molecule and is a gas at room temperature. As hydrocarbon molecules become more complex, they get heavier and become liquid at room temperature. Petroleum and diesel fuels are mixtures of several different types of hydrocarbon molecules. The carbon mass fraction of typical automotive fuels is approximately 0.86 to 0.87. This mass fraction approximates to a fuel that contains two hydrogen atoms for every one carbon atom (CH_2).

Thus, the mass of carbon is $1 \times 12 = 12$, and the mass of hydrogen is $2 \times 1 = 2$. The mass fraction of carbon is $12/(12 + 2) = 0.857$. Therefore, methylenecyclopropene (CH_2) is a useful approximation of the molar carbon-to-hydrogen ratio of a typical fuel when performing air/fuel ratio (AFR) calculations.

Since 1980, particulate emissions have reduced by two-thirds, but the levels are still considered problematic. Now the focus is on reducing levels of particulates with a diameter of less than 2.5 microns. Measuring likewise is problematic because particles of this size do not have a measurable mass. The favored control solution is via traps or by tackling the problem at the source by increasing combustion efficiency through advances in technology. A key element in particulate production is the sulfur content, and this was reduced to 50 ppm by 2005 in Europe, the United States, and the Asia Pacific Nations. In many countries, the legislated level is sub-5 ppm.

What Are Diesel Emissions?

Diesel engines convert the chemical energy contained in fuel into mechanical power. Diesel fuel is injected under pressure into the engine cylinder, where it mixes with air and where combustion occurs. The exhaust gases, which are discharged from the engine, contain several constituents that are harmful to health and to the environment.

Table 7.1 lists typical output ranges of the basic toxic material in diesel fumes. The lower values can be found in new, clean diesel engines; the higher values are characteristic of the previous generation of engine designs.

TABLE 7.1
EMISSIONS FROM DIESEL ENGINES

CO	HC	DPM	NOx	SO_2
vppm	vppm	g/m^3	vppm	vppm
5–1500	20–400	0.1–0.25	50–2500	10–150

vppm = volumetric parts per million.

Emission Constituents

CO (Carbon monoxide)	Mechanism:	Insufficient oxygen (a rich mixture) to fully oxidize the fuel
	Main controller:	Equivalence ratio
NOx (Nitrogen oxides)	Mechanism:	At high temperatures and in the presence of oxygen, nitrogen molecules react to form NO and NO_2
	Main controller:	Combustion temperature and oxygen availability
HC (Hydrocarbons)	Mechanism:	Some fuel escapes combustion (typically 1 or 2%)

	Main controller:	• Crevices
		• Incomplete combustion
		• Design issues
		  – Heterogeneous air/fuel mixture
		  – Cylinder-to-cylinder variation
		• Secondary issues
		  – Quenching
		  – Post-oxidation by hot exhaust
		  – Absorption and desorption from the oil layer
		• Abnormal sources
		  – Misfires
		  – Injection timing
		  – Exhaust valve leakage
		  – Valve guide leakage
		  – Piston ring sealing

Carbon monoxide (CO), hydrocarbons (HC), and aldehydes are generated as the result of the incomplete combustion of fuel. Significant portions of exhaust hydrocarbons also are derived from the engine lube oil. When engines operate in enclosed spaces, such as in closed garages, carbon monoxide can accumulate in the ambient atmosphere and can cause headaches, dizziness, and lethargy. Under the same conditions, hydrocarbons and aldehydes cause eye irritation and choking sensations. Hydrocarbons and aldehydes are major contributors to the characteristic diesel smell. Hydrocarbons also have a negative effect on the environment, being an important component of smog.

Nitrogen oxides (NOx) are generated from nitrogen and oxygen under the high-pressure and high-temperature conditions in the engine cylinder. Nitrogen oxides consist mostly of nitric oxide (NO) and a small fraction of nitrogen dioxide (NO_2). Nitrogen dioxide is very toxic. Furthermore, NOx emissions are a serious environmental concern because of their role in the formation of smog.

Sulfur dioxide (SO_2) is generated from the sulfur present in diesel fuel. The concentration of SO_2 in the exhaust gas depends on the sulfur content of the fuel. Low sulfur fuels of less than 0.05% sulfur are being introduced for most diesel fuel outlets throughout North America. Sulfur dioxide is a colorless toxic gas with a characteristic, irritating odor. Oxidation of sulfur dioxide produces sulfur trioxide, which is the precursor of sulfuric acid. In turn, sulfuric acid is responsible for sulfate particulate matter emissions. Sulfur oxides have a profound impact on the environment and are the major cause of acid rain.

Diesel particulate matter (DPM), as defined by U.S. Environmental Protection Agency (EPA) regulations and sampling procedures, is a complex aggregate of solid and liquid material. Its origin is carbonaceous particles generated in the engine cylinder during combustion. The primary carbon particles form larger agglomerates and combine with several other components, both organic and inorganic, of diesel exhaust. Generally, DPM is divided into three basic fractions:

1. Solids—Dry carbon particles, commonly known as soot

2. Soluble organic fraction (SOF)—Heavy hydrocarbons absorbed and condensed on the carbon particles

3. SO_4—Sulfate fraction, hydrated sulfuric acid

The actual composition of DPM will depend on the particular engine and its load and speed conditions. Wet particulates can contain up to 60% of the hydrocarbon fraction (SOF), whereas dry particulates comprise mostly dry carbon. The amount of sulfates is related directly to the sulfur contents of the diesel fuel.

Diesel particulates are very fine. The primary (nuclei) carbon particles have a diameter of 0.01~0.08 micron, whereas the diameter of the agglomerated particles is in the range of 0.08 to 1.0 micron. As such, diesel particulate matter is almost totally respirable and has a significant health impact on humans. It has been classified by several international government agencies as either a "human carcinogen" or a "probable human carcinogen." It also is known to increase the risk of heart and respiratory diseases.

Polynuclear aromatic hydrocarbons (PAH) are hydrocarbons containing two or more benzene rings. Many compounds in this class are known human carcinogens. PAHs in the exhaust gas are split between the gas and particulate phases. The most harmful compounds of four and five rings are present in the SOF of DPM.

How Are Emissions Regulated?

Regulations related to emissions and air quality may be divided into two classes:

- Tailpipe emission regulations
- Ambient air quality standards

All diesel engines for highway applications and some for off-road use are subject to the tailpipe emission regulations. These regulations specify the maximum amount of pollutants allowed in exhaust gases from a diesel engine. The emissions are measured over an engine test cycle, which also is specified in the regulations. The duty to comply is on the equipment (engine) manufacturer. All equipment must be emission certified before it can be released to the market. The authorities regulating engine tailpipe emissions include the EPA and the California Air Resources Board (CARB). Many applications of diesel engines in confined spaces are regulated through ambient air quality standards rather than by tailpipe regulations.

How Can We Control Diesel Emissions?

Diesel emissions are controlled either at their source, through engine design and modifications, or by exhaust gas after-treatment. The two approaches are complementary and are followed simultaneously in real life. There are two groups of diesel exhaust after-treatment devices:

1. Diesel traps
2. Diesel catalysts

Diesel traps, which are primarily diesel filters, control diesel particulate matter emissions by physically trapping particulates. The major challenge in the design of a diesel filter system is to regenerate the trap from collected particulate matter in a reliable and cost-effective manner. So far, diesel filters are used commercially in only a few specialized diesel engine applications (e.g., Peugeot).

Diesel catalysts control emissions by promoting chemical changes in the exhaust gas. They are most effective on gaseous emissions (i.e., hydrocarbons and carbon monoxide). Modern diesel catalysts also are becoming increasingly effective in controlling diesel particulate matter. Diesel catalysts have been used commercially for many on-highway and off-highway applications

What Is the Diesel Oxidation Catalyst?

Modern catalytic converters consist of a monolith honeycomb substrate coated with platinum group metal catalyst, packaged in a stainless steel container. The honeycomb structure, with many small parallel channels, presents a high catalytic contact area to exhaust gases. As the hot gases contact the catalyst, several exhaust pollutants are converted into harmless substances: carbon dioxide and water. The diesel oxidation catalyst is designed to oxidize carbon monoxide, gas phase hydrocarbons, and the SOF of diesel particulate matter to CO_2 and H_2O:

$$CO + | O_2 \rightarrow O_2$$

$$[Hydrocarbons] + O_2 \rightarrow O_2 + H_2O$$

$$[SOF] + O_2 \rightarrow CO_2 + H_2O$$

Diesel exhaust contains sufficient amounts of oxygen necessary for these reactions. The concentration of O_2 in the exhaust gases from diesel engines varies between 3 and 17%, depending on the engine load. Figure 7.2 shows typical conversion efficiencies for CO and HC in the diesel catalyst. The catalyst activity increases with temperature. A minimum exhaust temperature of approximately 200°C is required for catalyst light off. At elevated temperatures, conversions depend on the catalyst size and design and can be higher than 90%.

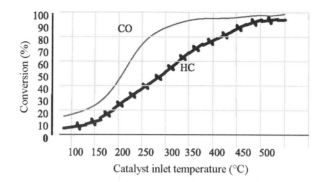

Figure 7.2 Catalytic conversion of carbon monoxide and hydrocarbons.

Conversion of diesel particulate matter is an important function of the modern diesel oxidation catalyst. The catalyst exhibits a very high activity in the oxidation of the organic fraction (SOF) of diesel particulates. Conversion of SOF may reach and exceed 80%. At lower temperatures (e.g., 300°C), the total DPM conversion usually is between 30 and 50% (Figure 7.3). At high temperatures (e.g., above 400°C), a counterproductive process may occur in the catalyst. The oxidation of sulfur dioxide to sulfur trioxide, which combines with water, forms sulfuric acid:

$$H_2O$$

$$SO_2 + \tfrac{1}{2}O_2 \rightarrow SO_3 \rightarrow H_2SO_4$$

Formation of sulfate (SO_4) particulates occurs, outweighing the benefit of the SOF reduction. Figure 7.3 shows an example situation, where at 450°C, the engine-out and the catalyst total DPM emissions are equal. In reality, the generation of sulfates strongly depends on the sulfur content of the fuel, as well as on the catalyst formulation. It is possible to decrease DPM emissions with a catalyst even at high temperatures, provided suitable catalyst formulation and good-quality fuels of low-sulfur content are used. On the other hand, diesel oxidation catalysts used with high-sulfur fuel will increase the

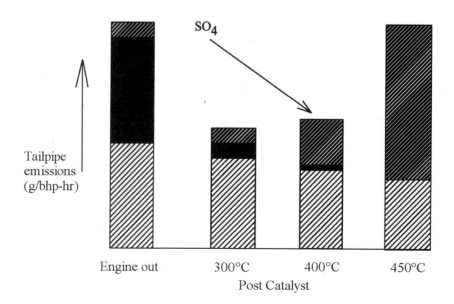

Figure 7.3 Formation of sulfate (SO_4).

total DPM output at higher temperatures. This is why diesel catalysts became more widespread only after the commercial introduction of low-sulfur diesel fuel.

Depending on its formulation, the diesel oxidation catalyst also may exhibit some limited activity toward the reduction of nitrogen oxides in diesel exhaust. NOx conversions of 10–20% usually are observed. The NOx conversion exhibits a maximum at medium temperatures of approximately 300°C.

What Are Hydrocarbon Traps?

Diesel engines are characterized by relatively low exhaust gas temperatures. When diesel engines operate at idle or with low engine load, the catalyst temperature may be lower than required for the catalytic conversion. At such conditions, the exhaust pollutants may pass untreated through the catalytic converter. A new diesel catalyst technology has been developed to enhance the low-temperature performance of the diesel oxidation catalyst. The technology incorporates hydrocarbon-trapping materials (HC traps) into the catalyst wash coat. Zeolites, also known as molecular sieves, are most frequently used as the hydrocarbon traps. These zeolites trap and store diesel exhaust hydrocarbons during periods of low exhaust temperatures, such as during engine idling. When the exhaust temperature increases, the hydrocarbons are released from the wash coat and are oxidized on the catalyst. Due to this hydrocarbon trapping mechanism, the catalyst exhibits low hydrocarbon light-off temperatures and excellent diesel odor control (Figure 7.4).

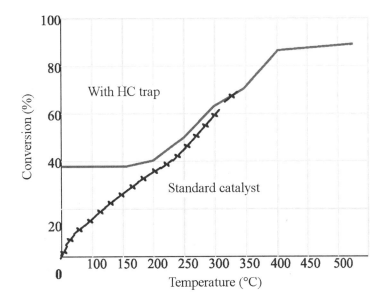

Figure 7.4 Hydrocarbon conversion in a catalyst with an HC trap.

The HC trapping catalysts are designed to work at transient engine conditions. Because the low-temperature performance occurs through adsorption rather than through catalytic conversion, periods of hot exhaust temperatures are needed for hydrocarbon desorption and regeneration of the catalyst. Otherwise, the adsorption capacity will become saturated, and increasing hydrocarbon emissions will break through the catalyst.

Additives for Regeneration of Fuel-Borne Catalysts

These are substances added to the fuel in tiny concentrations (10–20 ppm), thereby decreasing the incendiary temperature of soot due to the catalytic effect down to approximately 300–400°C. Examples of such substances are the elements cerium, copper, iron, and strontium. Final residues of these additives appear again in the exhaust as extremely small particles of ash (approximately 20 nm). Therefore, additives are admissible only in combination with suitable particle filters. No doubt the potential of additives is advantageous in substantially reducing the primary emissions of soot, which leads to lesser loads on the filters.

Catalytic Coatings

A similar method of decreasing the ignition temperature of soot is achieved by means of layers of transition metals in the filter, provided the specific surface of the filter material is rather large (>100 m^2/g), which is required for a very fine distribution of active centers of the reaction. When using coated filters, the formation of thick soot cakes should be avoided. While additives permit even burning of massive deposits of soot, albeit at high-temperature peaks, this would substantially impair the effects of the catalyst layer on the wall.

The Continuously Regenerating Trap System

The continuously regenerating trap (CRT) system uses a coated precious metal oxidation-catalyst prior to the trap to generate NO_2 from NO. However, NO_2 at the prevailing temperatures is unstable; therefore, in the particle trap afterwards, the reaction is reversed, and the free radical of oxygen oxidizes the carbon already at temperatures from approximately 250°C upward (Figure 7.5).

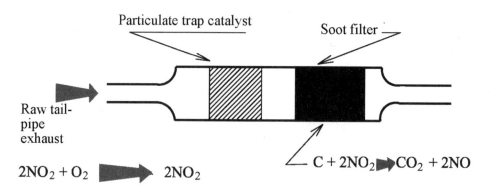

Figure 7.5 Example of a particulate trap.

The principal functions of the CRT system are as follows:

- Trapping of soot carbon
- Low-temperature regeneration
- Oxidation of soot carbon
- Oxidation of hydrocarbons from unburned fuel and oil

The CRT system requires sulfur-free fuel to avoid the sulfate reaction ($SO_2 \Rightarrow SO_3$) that otherwise would interfere with NO_2 conversion. Many variations of this method have been made. That of Peugeot developed for passenger cars may serve as an example in the following.

Cerium-oxide is used as an additive to the fuel for decreasing the ignition temperature of the soot by approximately 200°C. This is not sufficient in the case of passenger cars. In these cases, the temperature of the exhaust is increased a further 100°C by means of after-injection of fuel (Figure 7.6).

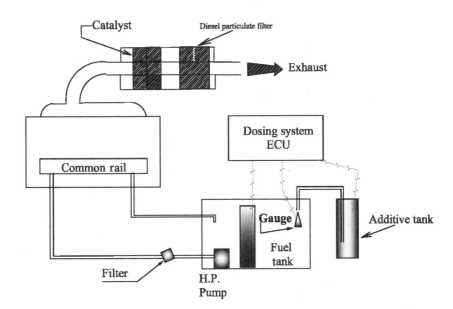

Figure 7.6 Fuel-borne catalyst system.

Incompletely burned fuel is brought to reaction in a pre-catalyst, which leads to a further increase of temperature. Regeneration of exhaust, serving for cooling the combustion, is interrupted during regeneration, and all electric devices are set under load, simply to increase the load on the engine. All these tricks are required to trigger regeneration under unfavorable conditions of low load, while the limit of admissible backpressure is reached. Filters must be cleansed of inert substances from time to time at intervals of approximately 2,000 operating hours, or about every 100,000 kilometers. The inert substances are caused primarily by additives in lubrication oil. There are metal-oxides for inhibition of wear, such as zinc and calcium, which are the principal components of anti-corrosion additives. Together with sulfur from fuel or lube oil, the inert substances may form gypsum, which is not welcome if deposited in the particle filter because it clogs the pores. This cleansing requires dismounting of the filter and careful flushing to dispose of the accumulated ash.

European emission regulations for new light-duty vehicles (cars and light commercial vehicles) originally were specified in the European Directive 70/220/EEC. Amendments to that regulation include the Euro 1/2 standards, covered under Directive 93/59/EC, and the Euro 3/4 limits (2000/2005), covered by Directive 98/69/EC. The 2000/2005 standards were accompanied by the introduction of more stringent fuel quality rules that required a minimum diesel cetane number of 51 (year 2000); a maximum diesel sulfur content of 350 ppm in the year 2000 and 50 ppm in the year 2005; and a maximum gasoline sulfur content of 150 ppm in the year 2000 and 50 ppm in the year 2005. The emission test cycle for these regulations is the ECE 15 + EUDC procedure. Effective in the year 2000, that test procedure is modified to eliminate the 40-second engine warm-up period before the beginning of emission sampling. All emission limits are expressed in grams per kilometer (g/km).

The EU light-duty vehicle standards are different for diesel and gasoline vehicles. Diesels have lower CO standards but higher NOx. Gasoline vehicles are exempted from particulate mass (PM) standards. Internationally year-on-year standards are becoming ever more stringent. The year 2010 will demand significant reductions, and targets for near-zero emissions by 2020 are being discussed.

What Makes Up Air?

Air is a mixture of oxygen molecules and nitrogen molecules, plus a small amount of other gases (Table 7.2). Oxygen and nitrogen molecules are diatomic; two oxygen atoms are bonded together to form an oxygen molecule, and two nitrogen atoms are bonded together to form a nitrogen molecule. The molecules move freely because air is a gas at room temperature.

TABLE 7.2
PRIMARY COMPONENTS OF AIR

Component	% by Volume	% by Mass
Nitrogen	78	75.5
Oxygen	21	23.1
Argon	0.9	1.3
Carbon dioxide	0.03	0.05

Oxygen molecules have a relative mass of 32; nitrogen molecules have a relative mass of 28. The mass fraction of oxygen is approximately 20%; the mass fraction of nitrogen is approximately 80%.

How Does Combustion Occur?

At room temperature, gasoline or diesel will not combust spontaneously in air, nor will either react with air. The fuel and the air are stable. Two things must occur for fuel and air to react:

1. The fuel and air must be mixed together in the correct proportions.
2. Some energy must be added to start the reaction.

In a gasoline engine, a spark is used to start combustion. In a diesel engine, the heat generated by the compression of the air trapped in the cylinder is used to start combustion (Table 7.3). Note that there are more than 10 million known organic compounds.

Atoms generally do not exist as single entities; they normally associate together to form molecules. Atoms associate with other atoms in many different ways. One way that is of interest when studying combustion is called covalent bonding (Figure 7.7).

The number of electrons present in the outer shell of an atom is a good indicator of the reactivity of that atom. If the outer shell is completely full of electrons, then the atom

TABLE 7.3
INTERRELATIONSHIP BETWEEN CARBON AND HYDROGEN

Number of Carbon Atoms	Number of Hydrogen Atoms	Number of Molecules
1	4	1
2	6	1
3	8	1
4	10	2
5	12	3
6	14	5
7	16	9
8	18	18
9	20	35

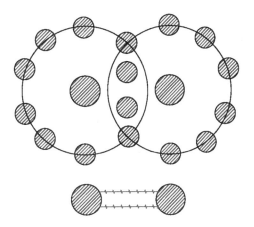

Figure 7.7 Covalent bonding.

usually is not very reactive. If atoms can fill their outer shells by sharing electrons with other atoms, then the molecule formed will be less reactive. When atoms are bonded together by sharing electrons, the bonding is called covalent bonding.

The greater the number of electrons that are shared, the stronger the bonding. Covalent bonds are described as single, double, or triple, depending on the number of shared electrons.

Consider a pair of balls that are attached by a spring and are in a state of constant agitation (Figure 7.8).

As energy is added to the air within the cylinder, the molecules move faster. The atoms within the molecules also oscillate relative to each other. Once the molecules have absorbed enough energy, the bonds between the atoms start to break.

The bonds between different types of atoms have different strengths or bond energies. The molecules containing atoms with the lowest bond energies will start to disassociate before the molecules containing atoms with higher bond energies (Figure 7.9).

Figure 7.8 *Molecule as a spring.*

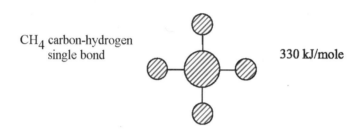

Figure 7.9 *Examples of some covalent bond energies.*

Consider the simple case of methane burning in air:

Step 1—Mix the fuel and air in the correct proportions (Figure 7.10).

Step 2—Add some energy to break the chemical bonds (i.e., compression leading to air temperature rise) (Figure 7.11).

Step 3—The molecules reform to CO_2 and H_2O (Figure 7.12).

Step 4—The energy given out as new molecules form causes other molecules to disassociate.

Combustion is moving through the mixture, until all the fuel has combusted, and surplus oxygen is giving rise to a weak mixture (Figure 7.13).

Note the bond energies of the products of combustion. The energy required to break the nitrogen molecule is nearly twice that of oxygen. High combustion temperatures liberate high energy levels, which will break the nitrogen triple bond and lead to the formation of NOx (Figure 7.14).

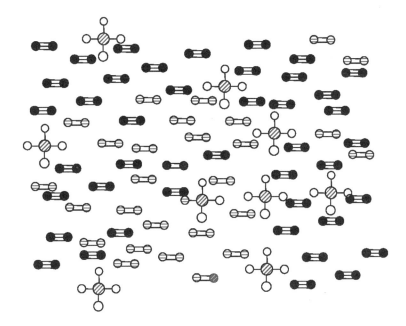

Figure 7.10 Methane molecules mixed with oxygen and nitrogen molecules.

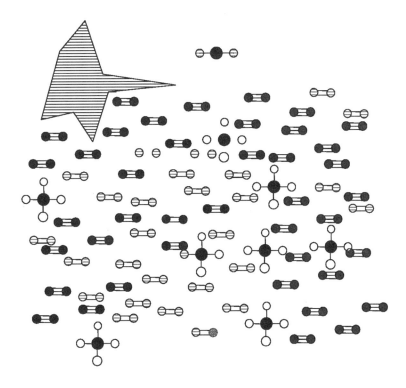

Figure 7.11 Ignition, combustion started. (Note some molecules with dis-associated atoms.)

Why Do Internal Combustion Engines Emit Gases?

When hydrocarbon fuel burns in air, some of the chemical energy contained in the fuel is converted into work. The nitrogen trapped in the cylinder is heated by the energy

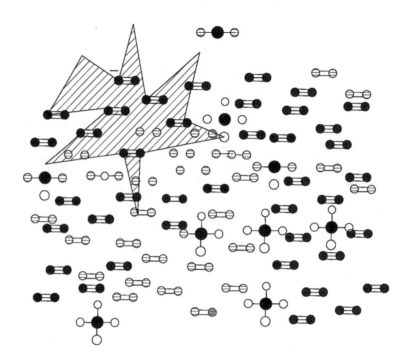

Figure 7.12 Some CO_2 and H_2O formed.

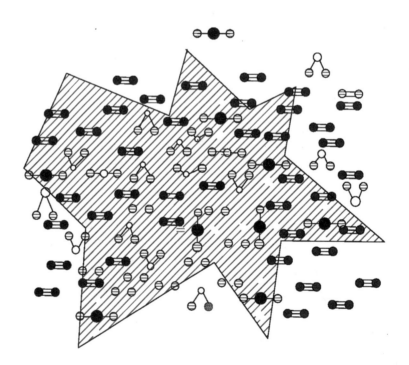

Figure 7.13 Combustion moving through the mixture.

released when the hydrogen and carbon in the fuel react with the oxygen in the air. When exhaust gases leave the cylinder, they are cooled by the surrounding air. The carbon dioxide gas produced will remain as a gas, but the steam produced will condense to form water droplets.

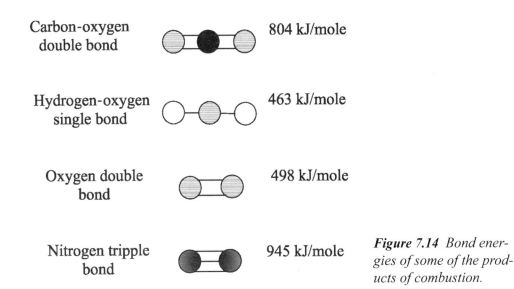

Figure 7.14 Bond energies of some of the products of combustion.

Balanced Chemical Equations

When one molecule of methane is burned in air, two molecules of oxygen are consumed. Air contains approximately eight molecules of nitrogen for every two molecules of oxygen. The products are one molecule of carbon dioxide and two molecules of steam. The eight molecules of nitrogen remain in their original state.

Therefore, a balanced chemical equation for the combustion of methane is

$$1CH_4 + 2O_2 + 8N_2 = 1CO_2 + 2H_2O + 8N_2$$

Calculation of the Air/Fuel Ratio Using Methane as the Simplest Fuel

The air/fuel ratio (AFR) is the ratio of the mass of the air to the mass of the fuel. For the complete combustion of methane, the AFR is

$$(2 \times 32) + (8 \times 28)/16 = 18$$

When all of the oxygen and all of the fuel are consumed in the combustion process, the AFR is said to be ideal.

Unfortunately, air does not contain oxygen and nitrogen by volume in the exact ratio of 2 to 8. The ratio is closer to 2.1 to 7.8, with other gases contributing 0.1. Therefore, a more accurate determination of the ideal AFR for methane is

$$1CH_4 + 2O_2 + 7.52N_2 = 1CO_2 + 2H_2O + 7.52N_2$$

The calculation of the AFR for methane is the mass of the air/mass of the fuel, as follows:

The mass of the fuel is $1C = 1 \times 12 = 12 + 4H = 4 \times 1 = 4$. Hence, the total mass of the fuel is 16.

The mass of the air is $2O_2 = 2 \times 32 = 64 + 7.52N_2 = 7.52 \times 28 = 210.56$. Hence, the total mass of the air is 274.56.

The AFR is the mass of air/mass of fuel. Thus, the AFR is $274.56/16 = 17.16$.

The steps to calculate the ideal AFR for a typical fuel (CH_2) are as follows:

Step 1—Write a balanced chemical equation for CH_2.

$$1CH_2 + 1.5O_2 + 5.64N_2 = 1CO_2 + 1H_2O + 5.64N_2$$

Step 2—Determine the mass of the fuel.

$$1C = 1 \times 12 = 12$$
$$2H = 2 \times 1 = 2$$

The total mass of the fuel is 14.

Step 3—Determine the mass of the air.

$$1.5O_2 = 1.5 \times 32 = 48$$

$$5.64N_2 = 5.64 \times 28 = 157.92$$

The total mass of the air is 205.92.

Step 4—Find the air/fuel ratio.

$$AFR = \text{Mass of air/mass of fuel}$$

$$AFR = 205.92/14 = 14.71$$

Dalton's Law of Partial Pressures

All molecules in a gas occupy the same amount of space; therefore, the percentage of carbon dioxide in the products of an engine running on a typical fuel can be found by the following:

Volume of the measured gas/volume of the whole exhaust gas

Using Dalton's Law

For a volume of exhaust gas, this can be expressed as the

Number of molecules of measured gas/total number of molecules of exhaust gas

The balanced chemical equation shows that the relative number of molecules of each product in the exhaust gas will exist only in a defined ratio. Therefore,

$$1(CO_2)/(1(CO_2) + 1.5(H_2O) + 5.64(N_2))$$

$$1/(1 + 1 + 5.64) = 13.1\%$$

This calculation is called a wet basis calculation because the steam (H_2O) produced by combustion has been included. If the exhaust gas is cooled before it is measured, the steam will condense, and the relative percentage of carbon dioxide present in the remaining exhaust gas will increase. The dry basis calculation is

$$1(CO_2)/(1(CO_2) + 5.64(N_2))$$

$$1/(1 + 5.64) = 15.1\%$$

If the concentration of oxygen, carbon dioxide, and carbon monoxide is measured in the exhaust gas, and the carbon mass fraction of the fuel is known, then the AFR can be determined from the balanced chemical equation or by using a look-up table or an AFR chart.

Non-Ideal Combustion—Formation of Pollutants

If more air is supplied than is required for complete combustion, some oxygen will be present in the exhaust gas. Consider the complete combustion of methane:

Ideal: $1CH_4 + 2O_2 + 8N_2 = 1CO_2 + 2H_2O + 8N_2$

Weak or lean: $1CH_4 + 3O_2 + 12N_2 = 1CO_2 + 2H_2O + 1O_2 + 12N_2$

If too little air is supplied, then CO_2 cannot be formed.

Rich mixture: $1CH_4 + 1.5O_2 + 6N_2 = 1CO + 2H_2O + 6N_2$

In a real engine, sufficient air normally is supplied; however, if the fuel is not mixed well enough with the air, there may be local areas of rich and lean combustion. In a real engine, it is usual to see small amounts of both CO and O_2 in the exhaust gases, together with some hydrocarbons that have survived the combustion process.

Fuel Droplets

The size of the fuel droplets must be minimized during combustion if ideal combustion is to be achieved (Figure 7.15). Any localized rich mixtures will give rise to high CO, HC, and particulate emissions. The rate of heat release, especially at the start of combustion, also is an important parameter in minimizing the formation of unwanted gases. If the rate of heat release is too high, then NOx will form. In common rail diesel applications, 2000-bar injection with very small droplet size poses a new problem. What is this problem?

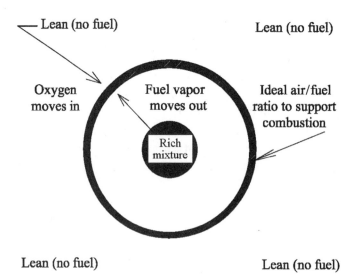

Figure 7.15 *A fuel droplet.*

The fuel droplets also must be kept away from the surfaces of the combustion chamber, cylinder walls, piston crown, and other crevices. Otherwise, complete combustion will not occur.

If the temperature of the combustion is too hot, oxides of nitrogen form.

Combustion Differences Between Diesel and Gasoline Engines

Traditionally, the AFRs of gasoline and diesel engines were very different. Gasoline engines were set up slightly rich to give optimum performance, whereas diesel engines were set up lean to reduce smoke emissions and to give good fuel economy (Figures 7.16 and 7.17). Where possible, most modern gasoline engines are set to cycle around the

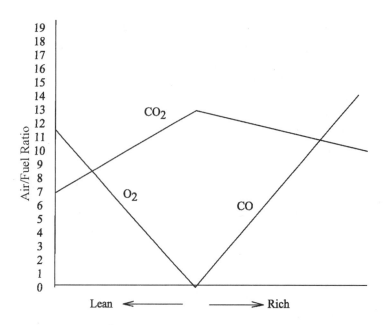

Figure 7.16 *AFR versus emissions.*

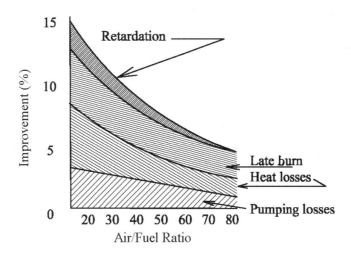

Figure 7.17 AFR versus retard effects.

ideal AFR to allow the catalyst to function correctly. A gasoline engine can run close to the ideal AFR without producing smoke because the fuel is more volatile.

Direct and Indirect Injection Diesel Engines

Comparisons between naturally aspirated direct injection (DI) and indirect injection (IDI) diesel engines of closely comparable design and size indicate that the DI engine is always more efficient, although the benefit varies with load (Figure 7.18). At full load, differences of up to 20% in brake specific fuel consumption (BSFC) have been noted, especially in engines with larger displacement per cylinder. At part load, the gain is less (circa 10%). Comparisons should be made with equal emission levels, a task that is difficult to accomplish in practice. Emission control with the DI engine is more difficult, so this constraint reduces the performance benefit. Figure 7.18 shows a breakdown of the indicated efficiency differences between the two systems. At full load, the IDI suffers a penalty of approximately 15 to 17%, due in part to the retarded timing of the IDI combustion process and its long, late-burning, heat release profile. At light load, approximately 300 kPa BMEP (brake mean effective pressure) (AFR 50:1),

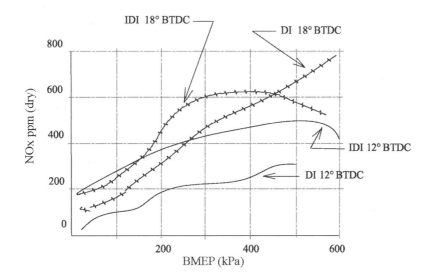

Figure 7.18 The effect of load on a naturally aspirated diesel engine.

the combustion effects are small, and the indicated efficiency penalty of the IDI (about 5 to 7%) is due to the higher heat losses associated with the larger surface area, and high-velocity flow through the connecting nozzle of the divided-chamber geometry, as well as due to pumping pressure loss between the main and auxiliary chambers. Figure 7.18 shows the effect of load on NOx emissions on a DI and IDI (pre-combustion chamber) six-cylinder 5.9-liter power unit.

At fixed speed and constant fuel delivery cycle, the DI engine shows an optimum BSFC and BMEP at a specific start of injection for a given injection duration. The IDI engine experiments are at a fixed BMEP; here, the BSFC at full load and fueling rate at idle shows a minimum at specific injection timings. Injection timing, which is more advanced than this optimum, results in combustion starting too late.

Figure 7.19 illustrates typical DI and IDI configurations.

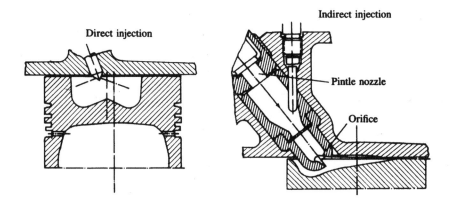

Figure 7.19 Typical DI and IDI configurations.

Injection timing is a factor that can improve the indicated efficiency of smoke and particulate emissions of naturally aspirated and turbocharged small DI diesel engines (Figures 7.20 and 7.21). The effect of the start of injection timing on diesel engine performance and emissions with a medium-swirl DI diesel engine with a deep combustion bowl and a four-hole injection nozzle at 2600 rev/min, fuel delivery 75 mm^3/cycle. In this example, the fuel/air equivalence ratio is 0.69.

Figures 7.20 and 7.21 refer to smoke (Bosch smoke number) and particulate mass emissions (in grams per kilowatt hour) as a function of load and injection timing for a six-cylinder 3.7-liter IDI swirl chamber diesel engine at 1600 rev/min. Note that there is no exhaust gas recirculation (EGR).

Exhaust gas recirculation, where a controlled mass of exhaust gas is fed into the induction system, is a popular method of reducing NOx levels. Uncontrolled EGR due to the flow reversals across the cylinder head between the induction and exhaust ports is thought to be a primary cause of unstable running in engines. Running at constant speed, note the effect of increasing the levels of EGR, as shown in Table 7.4.

A medium-capacity DI diesel engine running at 1500 rev/min and developing 88 Nm (Figure 7.22) illustrates the effect on brake specific hydrocarbons with differing injection timing and variable EGR. Figures 7.23 and 7.24 show various aspects and effects of EGR.

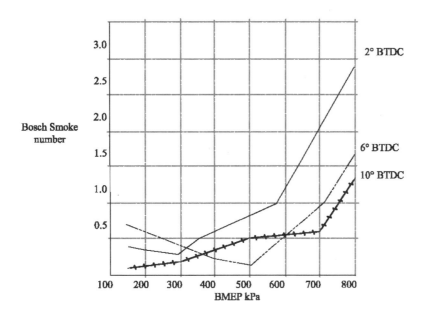

Figure 7.20 Effect of injection timing on smoke.

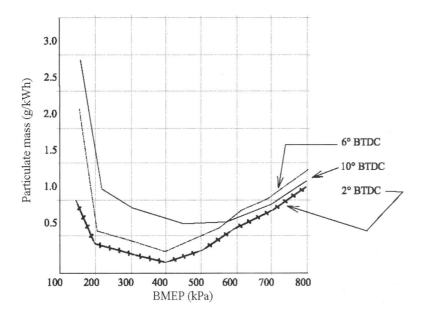

Figure 7.21 Effect of timing on particulates.

Note the minimal effect on emissions when the point of pilot injection is moved when running at constant speed (Figures 7.25 through 7.27).

Compare Figure 7.26 with Figure 7.28. This is a mirror image and is typical of the relationship between fuel used and BMEP/torque.

At part load, EGR can be used to reduce diesel engine NOx emissions. Note that because diesel engines operate with the airflow unthrottled, at part load the CO_2 and H_2O concentrations in the exhaust gas are low. They essentially are proportional to the fuel/air ratio. Because of this, high EGR levels are required for significant reductions in NOx emissions.

TABLE 7.4
EGR SWING FROM 0 TO 100% MODULATION

CO	CO$_2$	O$_2$	HC	NOx	FSN	%EGR
1760	12.1	3.8	118	60.6	3.53	100
1635	12	3.9	118	61.7	3.41	90
1458	11.8	4.3	118	66.2	3.08	80
1268	11.4	4.7	117.1	75.9	2.58	70
1080	10.9	5.4	110	94.8	1.96	60
930	10.3	6.3	101	122.7	1.22	50
875	9.4	7.5	91.5	184.4	0.42	40
815	7.9	9.6	74.5	343	0.08	30
800	6.9	11	67.4	512	0.04	20
880	6.5	11.6	70.6	577	0.05	10
865	6.7	11.3	59.1	541	2.1	40

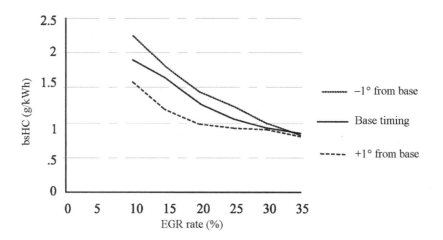

Figure 7.22 Effect of differing rates of EGR.

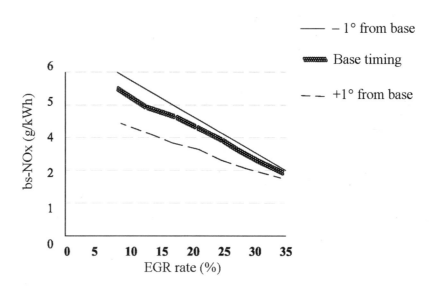

Figure 7.23 Effect of differing rates of EGR on NOx emissions when running at 1500 rev/min and developing 130 Nm.

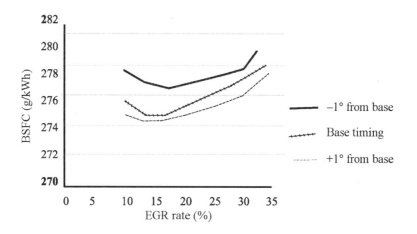

Figure 7.24 Fuel consumption versus EGR and injection timing.

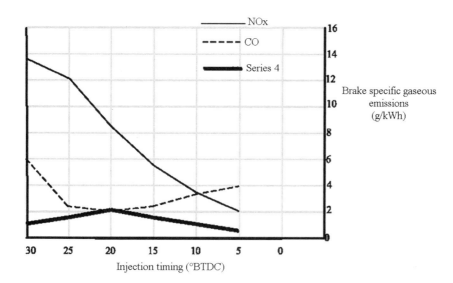

Figure 7.25 Medium swirl DI timing swing.

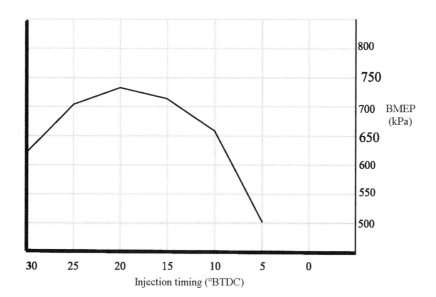

Figure 7.26 Timing swing BMEP loop.

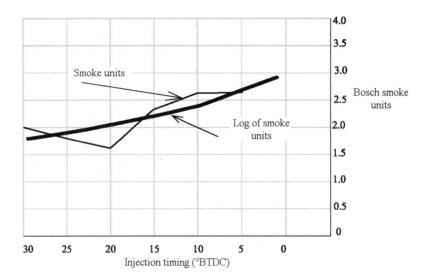

Figure 7.27 Effect on visible smoke when injection timing is changed.

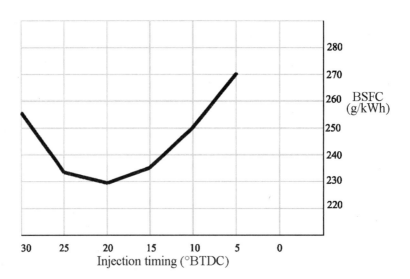

Figure 7.28 Fuel loop.

NOx concentrations decrease as the inlet airflow of a DI diesel engine is diluted at a constant fueling rate. The dilution is expressed in terms of oxygen concentration in the mixture after dilution. Table 7.4 shows how EGR affects specific NOx and HC fuel consumption, as well as smoke, for a small high-swirl DI diesel engine at a typical automobile part-load condition. Effective reduction of Bosch smoke NOx is achieved, with a modest reduction in brake specific hydrocarbon (bsHC) and only a small increase in BSFC. However, note that smoke increased as the rate of EGR increased.

Fuel injection timing essentially controls the crank angle at which combustion starts. Although the state of the air into which the fuel is injected changes as injection timing is varied and thus ignition delay will vary, these effects are predictable. The fuel injection rate, fuel nozzle design (e.g., number of holes), and fuel injection pressure all affect the characteristics of the diesel fuel spray and its mixing with air in the combustion chamber.

Injection timing variations have a strong effect on NOx emissions for DI engines; the effect is significant but less for IDI engines. Retarded injection commonly is used to help control NOx emissions. It gives substantial reductions initially, with only modest BSFC penalty. For the DI engine at high load, specific HC emissions are low and vary only modestly with injection timing. At lighter loads, HC emissions are higher and increase as injection becomes significantly retarded from optimum. This trend is especially pronounced at idle. For IDI diesel engines, HC emissions show the same trends but are much lower in magnitude than DI HC emissions.

Retarding timing generally increases smoke, although trends vary significantly among different types and designs of diesel engines. Mass particulate emissions increase as injection is retarded.

The higher injection rate depends on the fuel-injector nozzle area and injection pressure. Higher injection rates result in higher fuel/air mixing rates and hence higher heat release rates. For a given amount of fuel injected per cylinder per cycle, as the injection rate is increased, the optimum injection timing moves closer to top dead center (TDC). The higher heat release rates and shorter overall combustion process result from the increased injection rate and decrease the minimum BSFC at optimum injection timing. However, a limit to these benefits soon is reached. Increasing the injection rate increases NOx emissions and decreases smoke or particulate emissions. The controlling physical process is the rate of fuel/air mixing in the combustion chamber. Thus, at constant fuel injected per cylinder per cycle, both increased injection pressure at the fixed nozzle orifice area (which reduces the injection duration) and reduced nozzle area at the fixed injection duration produce these trends.

In a diesel engine, the mixing time is short because the fuel is injected toward the end of the compression stroke.

Operating Principles of Emission Reduction Devices Fitted to Internal Combustion Engines

Catalyst Operation

The CO and HC produced by non-ideal combustion can be oxidized to CO_2 by passing the exhaust gas over a catalyst. The exhaust gas must contain enough oxygen to complete the reaction of the CO and HC to CO_2. The purpose of the catalyst is to lower the energy input required to initiate and sustain the reaction. Although the activation energy is reduced, the catalyst must be hot to function correctly.

The catalyst is not consumed during the reaction process; however, the efficiency of the catalyst can be reduced by deposit buildup on the surface. The catalyst also can be damaged by excessive temperature, vibration, or physical impact.

If the oxygen content of the exhaust gas is cycled between rich and lean, reduction of NOx to N_2 and O_2 can be achieved in addition to the oxidation of CO and HC. This is called three-way conversion. Some of the O_2 produced during the reduction of NOx can be used to oxidize CO and HC emissions. Catalysts now are being produced that can oxidize particulate emissions (Figure 7.29).

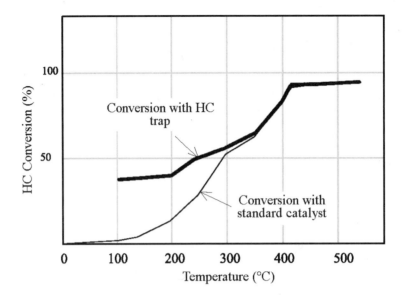

Figure 7.29 Hydrocarbon conversion in a catalyst with an HC trap.

Exhaust Gas Recirculation

Excessive combustion temperatures produce NOx emissions; unfortunately, it is difficult to reduce combustion temperatures without reducing the efficiency of the engine. During lean engine operation, there is excessive oxygen in the combustion chamber. If this oxygen is replaced with CO_2 and H_2O taken from the exhaust gas, then the opportunity for NOx to be created is reduced. Recirculating hot exhaust gas can increase the combustion temperature and can cause throttling of the inlet charge by reducing the mass flow of the intake air. Therefore, care must be taken not to recirculate exhaust gas because doing so may have a detrimental effect on the power output or may increase the levels of unwanted emissions.

Effects of Brake Mean Effective Pressure

As oxygen is used during the combustion process, the emission of CO, smoke, HC, and particulates normally rises because complete combustion to CO is not achieved. As more fuel is put into the combustion chamber to increase the power output, more oxygen is consumed; therefore, there is a link between power output and emissions. If an engine is operated in the rich range, the additional fuel in the combustion chamber will absorb energy as it changes phase. It also will cause the mass of the combustion gases to increase; therefore, the combustion temperature will fall. This reduction in temperature will cause the NOx emission to fall.

Principles and Operation of Raw, Dilute, Continuous, and Bag Sampling

Dilute and Raw Sampling

Dilute and raw sampling are two methods of obtaining exhaust gas for analysis. In dilute sampling, the blower draws air into the sampling system (Figure 7.30). The air then is filtered and mixed with the exhaust gas. Particles are separated from the air and exhaust gas by the cyclone. The flow rate of the mixed air and exhaust gas is regulated to a constant value by the critical flow venturi. A sample of the dilutant air is put into the ambient air bag. When the vehicle is operated over a drive cycle, a sample of the mixed air and exhaust gas is put into the sample bag. The flow rate of the samples is regulated to ensure that the samples are representative of the emissions produced during the drive cycle. The drive cycle may be split into several phases; there will be one ambient bag and one sample bag for each phase. At the end of the test, the contents of both bags are analyzed. The level of each pollutant is determined by subtracting the measured value in the ambient bag from the measured value in the sample bag.

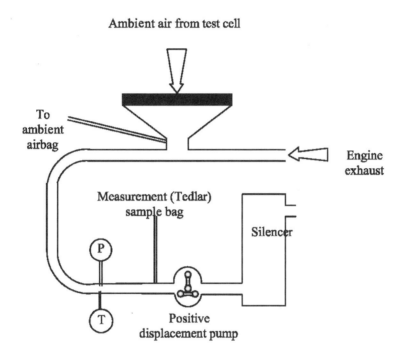

Figure 7.30 Basic CVS system.

Particulate Measurement

Before continuing, it is useful to give a definition of particulate. In the context of automotive engineering, it refers to the matter collected on a pair of fluorocarbon-coated glass fiber filter papers, or fluorocarbon-based membrane filter papers with a minimum stain diameter of 60 mm. The diluted exhaust gas temperature must be below 51.7°C (measured in the sampling zone), per U.S. EPA Regulation 40. Particulates consist of several materials. The main constituents are carbon, condensed and absorbed hydrocarbons, and sulfates. The particulate analyzer cannot determine the relative quantities of each constituent; it can only measure the combined mass of all of the constituents.

Principles and Operation of Flow Tunnels

To measure the particulate matter in the exhaust gas, all particulate matter must be removed from the dilutant air. The filtered dilutant air then is passed down a tunnel, where it is mixed with exhaust gas. To achieve good mixing, the flow rate must be sufficient to ensure that the air is turbulent. The exhaust gas is cooled by the dilutant air to simulate exhaust gas leaving the exhaust pipe of a vehicle. A sample of the cooled exhaust gas and dilutant air is drawn from the tunnel and is filtered. The particulates are trapped on the filter papers. The flow rate through the filter papers and the mass of particulate on the filter paper are measured. This gives the concentration of particulate matter per unit volume of diluted exhaust gas sampled (in grams per cubic meter). For example, if x grams of particulate are collected on a filter paper during a sample period of y seconds, and the flow rate of the sample is in cubic meters per second, then the concentration of particulate is

$$\text{Flow rate}/(y \times z)\left[g \times s \times m^{-3} \times s^{-1}\right] = \left[gm^{-3}\right]$$

where

- x = grams of particulate on filter paper
- y = sample time
- s = sample volume
- (gs) = total volume of diluted gas
- z = concentration pf particulate

The total particulate emission for the engine is calculated as follows. First, find the concentration of particulate in the sampled volume of gas, and then take the ratio of the sampled volume and the total volume of the diluted exhaust gas. Finally, multiply the two together.

Several features of the design of a particulate tunnel are important to achieve accurate and repeatable results. If a new tunnel is inspected after a few hours of operation, a significant amount of particulate will have been deposited on the walls of the tunnel and in the exhaust inlet or transfer tube. Any measurements taken during this period therefore will be light because some of the particulate that should have been deposited on the filter paper now is on the walls of the pipe-work and so forth. Therefore, the tunnel must be conditioned before accurate measurements can be taken.

If the tunnel is disturbed, some of the particulate deposits may be transported from the pipe-work and onto the filter paper. This will make a measurement heavy. To test the stability of a tunnel, it is useful to run some dummy tests with no exhaust gas flow before the tunnel is used. If the mass of the filter papers does not change during the dummy test, then the tunnel can be assumed to be stable.

The temperature of the tunnel and the pipe-work should be kept as constant as possible. Changes in temperature can affect the rate at which particulate is deposited onto the walls and the pipe-work.

The temperature of the dilution air should be maintained to 25 ± 1°C. Although this recommendation is much more stringent than most regulations, it is important to maintain accurate control of air temperature if good repeatability is to be achieved.

The dilution ratio is the mass flow rate of the exhaust gas entering the tunnel divided by the mass flow rate of the dilutant air. A target dilution ratio of approximately 6:1 to 10:1 should be set for the tunnel. If high dilution ratios are used, then the sensitivity of the measurement is reduced. If low dilution ratios are used, then the nature of the particulates formed in the tunnel may become unstable. At all times, the dilution ratio should never be allowed to fall below 4:1. Figures 7.31 and 7.32 show examples of two commonly used types of flow tunnels.

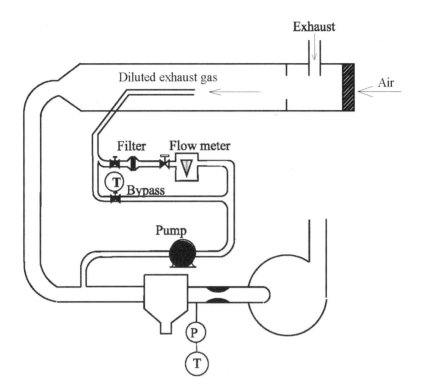

Figure 7.31 Full-flow particulate tunnel.

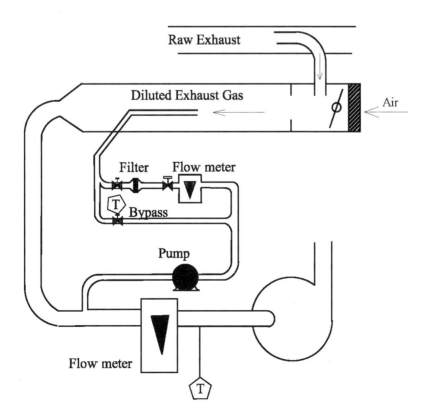

Figure 7.32 Simplified mini-particulate tunnel.

Filter Handling and Weighing

The amount of particulate deposited on the filter paper is small when compared to the weight of the paper. To achieve accurate results, it is important to handle the filter papers carefully. If any dust or dirt is allowed to get onto the paper, this will increase the mass of the paper and give rise to a heavy result. If any damage is caused to the paper (e.g., too high a clamping pressure in the filter holder), then the mass of the paper will be reduced, and a light result will occur. Figure 7.33 shows an example of a typical filter paper. The balance used to weigh the filter papers must be of sufficient accuracy to detect changes in the microgram range. Also, it must be fitted within a closed weighing chamber to reduce the effects of air currents. Calibration of the balance must be done before use and checked regularly with weights that are traceable to national standards. Two clean reference papers should be weighed before each batch of test filter papers. If there is any change in the mass of the filter papers, the precision of the balance and the humidity of the conditioning environment must be checked. The amount of water absorbed by the paper must be consistent during the pre- and post-test measurements. To achieve this, the paper must be conditioned in a clean humidity-controlled cabinet for several hours before weighing.

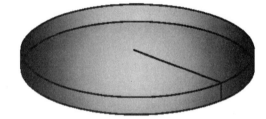

Figure 7.33 A 72-mm-diameter standard reference filter paper.

In conclusion, the major advantages of a full-flow tunnel are as follows:

1. The flow rate of the engine exhaust does not have to be measured.

2. The mixed gas flow rate is constant, even for transient test applications.

3. The constant volume control system is simple, relatively accurate, and generally reliable.

Principles and Operation of Micro-Tunnels

In a full-flow tunnel, only a tiny proportion of the mixed gas is drawn through the filter paper. If a fixed proportion of exhaust gas is diverted into the dilution tunnel (rather than taking all of the exhaust gas), then a much smaller tunnel and dilution system can be employed. The amount of particulate emission measured in the sampled gas then can be multiplied by the proportion of sampled gas to total exhaust gas to determine the total particulate emission for the engine. The micro-tunnel operates by first measuring or estimating the exhaust gas flow rate. A controlled quantity of dilution air is mixed with the exhaust gas to create a similar dilution ratio to that which would have occurred within a full-flow tunnel. The whole of the mixed gas then is passed through the filter papers where the particulate is extracted. The flow rate of the dilution air is adjusted rapidly to follow the changes in exhaust gas flow caused by changes to engine operating conditions. The quantity of exhaust gas sampled is calculated by subtracting the measured amount of dilution air added to the sampled exhaust gas from the measured total flow of mixed gas.

Principles and Operation of Mini-Tunnels

Mini-tunnels normally are suitable only for steady-state testing. This is true because the gas splitting system is not able to respond in sufficient time to follow the rapid changes in exhaust gas flow rate encountered when running transient tests. Several types of gas splitting systems are in use; a common one uses an isokinetic probe to maintain a fixed ratio of sampled gas flow to exhaust gas flow. The isokinetic probe functions by keeping the average velocity of the sampled gas in the probe equal to the average velocity of the gas in the exhaust pipe. If equal velocities are maintained, then the sampled gas flow rate is proportional to the total gas flow rate. The ratio of the amount of gas sampled to the total exhaust gas flow then is equal to the ratio of the cross-sectional area of the probe to the cross-sectional area of the exhaust pipe. The velocity of the gas in the probe is controlled by equalizing the static pressure of the sampled gas in the mouth of the probe, with the static pressure of the gas in the exhaust pipe (measured in alignment with the tip of the probe). Mini-tunnels can function adequately if the isokinetic probe is kept clean and the temperature changes of the surface of the transfer tube can be controlled or at least made to be repeatable from test to test.

The two major advantages of mini-tunnels and micro-tunnels are as follows:

1. They require less space.

2. The dilution air can easily be filtered and conditioned; this will become a major advantage as the amount of emissions from engines continues to fall to very low levels.

Principle of Continuous Particulate Analyzers

If a continuous measurement of engine particulate emission is required, then the filter paper technique is not suitable. Several devices can be attached to a conventional full or partial flow tunnel to measure continuous particulate emission. One technique now in common use is the tapered element oscillating microbalance (TEOM), as shown in Figure 7.34.

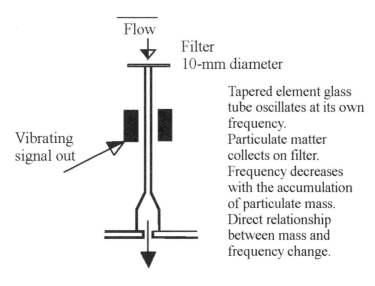

Figure 7.34 Tapered element oscillating microbalance (TEOM).

In this device, the particulate filter paper is attached to a hollow tube. A vacuum pump draws a controlled amount of diluted exhaust gas through the filter paper. A constant force is applied to the tube, which causes it to vibrate. As particulate builds on the filter paper, the increased mass causes the frequency of the vibration to change. This change in frequency is detected and converted into a measurement of continuous particulate mass emission.

The bag method of sampling returns one value for each pollutant for each pair of bags. If the level of pollutants for the transient parts of a drive cycle is required, then the sample must be measured continuously. If catalyst performance is to be measured, then two continuous raw samples must be taken: one upstream of the catalyst, and one downstream of the catalyst (or three if mid-bed values are required). If measurement of EGR is required, then a fourth raw sample must be taken from the intake manifold.

Hot and Cold Sampling

To prevent hydrocarbons from condensing on the walls of the cold raw sample lines, the lines that feed the hydrocarbon (and often the NOx) analyzer are heated to 190°C. Therefore, the total number of raw sample lines on a typical system is five (hot and cold pre-catalysts, hot and cold post-catalysts, and EGR) but could be seven if mid-bed catalyst measurements are required.

Essential Elements of a Sampling System

Whatever sampling system is being used, it is crucial that certain criteria are met. The exhaust gas must be presented to the analyzers in the required condition. The gas must be at the correct temperature, pressure, flow rate, and dryness and must be free from solid particles. The sampling system must not allow any air to enter. The materials must be compatible with the gas being sampled, and the material must be resistant to attack from the gases or any substances formed in the sampling system. Furthermore, the materials must not alter the composition of the sampled gases. The sampling lines must be able to be flushed with clean air. The sampling system must not alter the concentrations of the exhaust gases produced by the engine, and it must be non-intrusive. The sampling system must not trap gases (particularly hydrocarbons) and release them at a different time.

Temperature Control

Heated lines are used to control the temperature of the sample gas. The temperature must be hot enough to stop water and hydrocarbons from condensing in the line. The line should not have any hot or cold spots.

Pressure Control

Many analyzers are sensitive to changes in sample pressure and flow rate. The sample gas must be presented to the analyzers at the same pressure and flow rate as the span and zero gases. Therefore, the sampling system must be able to compensate for any change in sample pressure or flow rate caused by changes to the pressure at the exhaust sample point, or the changes due to filters blocking during sampling. The normal method of control is to bypass a significant quantity of gas via a backpressure regulator. The sample pressure and flow to the analyzers are kept constant by varying the flow rate of the bypass gas.

Flow Rate

The time taken for the sample gas to pass from the vehicle exhaust to the analyzer must be kept constant to time-align the continuous results to the drive cycle. The high bypass flow rate helps to shorten the response time and to lessen the time-alignment error due to partial filter blockage.

Dryness

Water vapor can cause measurement errors by condensing on optical lenses, absorbing infrared radiation, and reducing the concentration of some gases by absorption. Water vapor in the sample normally is reduced to a low level by passing the sample gas through a coil cooled to 4°C. The water usually is collected in a reservoir at the base of the coil and then is blown out periodically.

Solid Particles

Solid particles can cause measurement errors by blocking capillaries and absorbing infrared radiation. A filter usually removes the solid particles. The filter normally is the first component in the sampling system after the connecting line in order to protect as much of the sampling system as possible.

Sample Pump

The sample pump must be able to generate a sufficient and stable flow rate, the wetted parts must be resistant to attack from the sample gas, and it must not add to, or lose, any of the components being measured.

Leak Checking

Sampling additional air from the surroundings probably is the most common form of measurement error. Small leaks are difficult to detect because the exhaust gas usually contains some oxygen. Leaks are common because the joints in the sample line often are disturbed (i.e., when connecting a sample line to a new vehicle or when changing a filter). Leaks can be found by overflow checking the sampling system with nitrogen or a cocktail mixture of several reference gases (e.g., NOx, HC, and CO).

Back Flushing

To prevent the sampling system from becoming contaminated too quickly, it is important to back flush the system between readings. The back flush blows contaminants back down the sample line and into the engine exhaust pipe. It also drains any water that has been trapped below the cooling coil.

Non-Dispersive Infrared Analyzer

A gas will absorb light energy of a frequency band that generally is peculiar to the gas. If this frequency band is narrow and is not shared by any of the other gases that are likely to be present in the sample, then the amount of energy within a certain frequency band that is absorbed by a sample of exhaust gas can be determined. The amount of energy absorbed is proportional to the concentration of the gas to be measured. The non-dispersive infrared (NDIR) analyzer works by passing a pulsed beam of infrared light through a chamber containing the sampled gas. The amount of light absorbed by the measured gas is detected and is converted into an electrical signal. The electrical signal then is scaled to reflect the concentration of the measured gas. The light is pulsed by a chopper, which is a rotating disc with openings that alternately stop and allow the light to pass. The light normally is passed through a chamber containing nitrogen, called the reference cell, and the chamber containing the gas to be measured is called the sample cell. The detector contains the same gas as the gas to be measured. The detector measures the difference in the amount of light energy absorbed between the gas in the sample cell and the gas in the reference cell. Figures 7.35 and 7.36 show the basic principle of an NDIR analyzer and infrared absorption wavelengths.

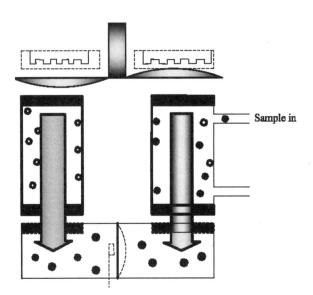

Figure 7.35 Basic principle of an NDIR analyzer.

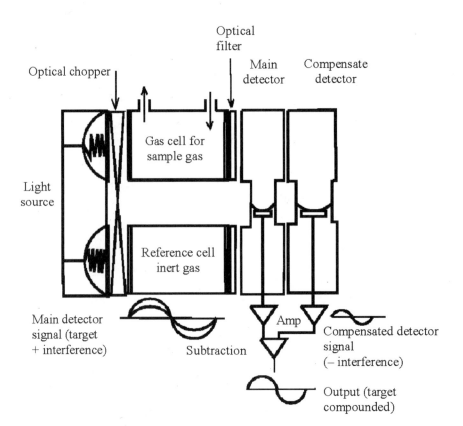

Figure 7.36 Horiba cell; typical reference cell NDIR CO and CO_2.

Flame Ionization Detector

In the flame ionization detector (FID), a sample of exhaust gas is passed through a flame. The flame burns any hydrocarbons present in the sample gas. When an electrical current is passed across the flame, the current changes in proportion to the amount of hydrocarbons contained in the sample gas. The electrical signal from the FID then is amplified and is passed to a display and logging device. The flame is sustained by mixing a gaseous fuel and air within a burner. The sample gas is introduced to the flame by mixing it with the fuel. Maintaining stable and precise flow rates of the sample gas, burner fuel, and burner air is vital if good accuracy is to be achieved. The flow rate of each gas is set by controlling the pressure drop across a capillary. Therefore, maintaining a constant pressure (especially between the sample and span gases) and a clean capillary are vital to achieve good performance. Care also must be taken to prevent any unwanted hydrocarbons from entering the burner, sampling system, or fuel and air lines. Contamination will show up as a high background level reading, and a true zero will be impossible to achieve. For this reason, high-purity gases must be used for the burner fuel and air. The basic principle of an FID analyzer is demonstrated in Figure 7.37, and an FID is shown in Figure 7.38.

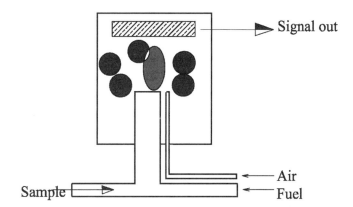

Figure 7.37 Basic principle of an FID analyzer.

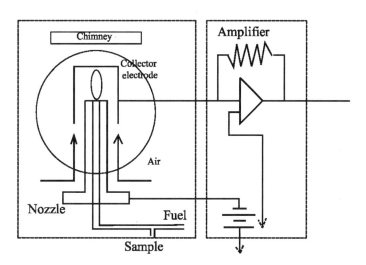

Figure 7.38 A flame ionization detector (FID) total hydrocarbon (THC) measurement schematic. Ionize HC molecules in a hydrogen flame.

NOx Analyzers

NOx (chemiluminescence) analyzers measure NO by detecting the light emitted when NO is reacted with ozone (O_3). The intensity of the light is proportional to the amount of NO in the sample gas. The intensity of the light is measured and converted to an electrical signal, mixed together in a reaction chamber. This chamber normally is evacuated by a vacuum pump to improve the sensitivity and stability of the analyzer. Other oxides of nitrogen are measured by passing them through a converter located before the reaction chamber. The converter changes any NO_2, NO_3, and so forth present in the sample gas into NO. The converter can be bypassed so that either the total NOx or NO alone can be measured. The analyzer relies on the precise control of the flow rate of the sample gas and ozone. The flow rates are maintained by controlling the pressure drop across the sample and ozone capillaries. The vacuum pump must give a stable vacuum to maintain a constant pressure drop across the capillaries. All of the ozone must be removed from the outlet of the reaction chamber to prevent a health hazard. The basic principle of a chemiluminescence NO/NOx analyzer is given in Figure 7.39, and a chemiluminescence detector is shown in Figure 7.40.

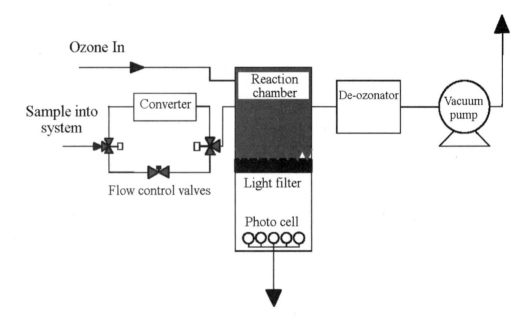

Figure 7.39 Operating principle of the chemiluminescence NO/NOx analyzer.

Oxygen Analyzers

At least three types of oxygen analyzers are in common use, as follows:

1. Those that work using the property of paramagnetism

2. Those that use oxygen as part of the electrolyte in an electrical cell

3. Those where the output of a fuel cell will vary, depending on the level of oxygen that enters the analyzer.

Paramagnetic analyzers have an advantage over the other two types because the sensing element does not have to be replaced regularly (as in the case of the fuel cell type)

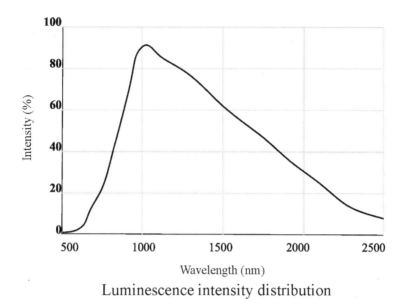

Luminescence intensity distribution

Figure 7.40 Chemiluminescence detector.

or serviced (as in the case of the electrolyte type). Figure 7.41 gives an example of a paramagnetic O_2 analyzer.

Oxygen is strongly paramagnetic, which means that the oxygen molecules will tend to align themselves in a way that adds to the strength of a magnetic field. In the design of one popular instrument, the exhaust gas is passed through a chamber shaped similarly to a dumbbell. The aligning force within the oxygen molecules is measured, and this force therefore is proportional to the amount of oxygen in the exhaust gas. Another popular method is to create a magnetic field around the outlet of a small pipe. As oxygen collects around the end of the pipe, it restricts the amount of nitrogen that will flow through the pipe. The change in flow of the nitrogen therefore is proportional to the amount of oxygen in the exhaust gas.

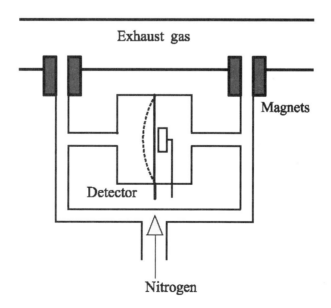

Figure 7.41 Paramagnetic O_2 analyzer.

Most gas analyzers measure the concentration of a gas by comparing the gas with two known reference points. The reference points are known as the zero and span values. The analyzer is calibrated before each measurement by passing a zero gas (usually nitrogen or air) through the analyzer and setting the response of the analyzer to zero. A gas containing a known concentration of the gas to be measured (span gas) then is passed through the analyzer, and the sensitivity or gain of the analyzer is adjusted to match the value of the span gas (Figure 7.42).

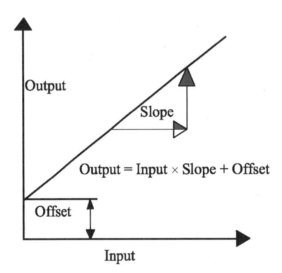

Figure 7.42 Calibration example showing $y = mx + c$ relationship for an analyzer, where c = zero error and m = gain.

The concentration of the zero and span gases must bracket the expected concentration of the sample gas. To choose the correct span gas, a small sample of exhaust gas should be taken before the analyzer finally is calibrated. The process of pre-sampling before calibrating is called sniffing.

Time Alignment

When continuous gaseous emission measurements are taken during transient engine or vehicle operation, it is important to be able to identify the correct time at which a particular level of emission was produced by the engine. Being able to correctly identify the time when a particular emission was produced is important for two reasons:

1. To be able to link the emission level with an event in the vehicle drive cycle or engine test program

2. To be able to calculate the real-time AFR

To time-align all of the gaseous emission analyzers, the time taken for the exhaust gas sample to travel to each analyzer and the concentration determined must be measured. When all data from the test run are processed, the emission data points must be repositioned versus test time by an amount equal to the measured time delay because each type of analyzer has different response times.

Maintenance

The secret to successful emission measurements is good housekeeping and attention to detail (e.g., keeping the sample lines clean and undamaged, changing the filters regularly, leak checking, calibrating before each measurement). Daily servicing and operating procedures should be established, together with regular routine maintenance and a system calibration plan (normally performed every three months).

Calibration of Analyzers Using a Gas Divider—NOx Efficiency Checks

The linearity of analyzers should be checked after any routine maintenance or major repair. The linearity is checked by accurately blending zero and span gases using a gas divider (Figure 7.43).

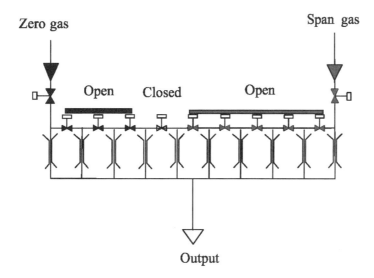

Figure 7.43 Signal instruments gas divider.

A NOx efficiency check is performed by diluting span gas with NO_2 and O_2. The amount of NO_2 in the span gas is measured by putting the NOx analyzer in NO mode and measuring the reduction in level of NO as NO_2 is introduced. The analyzer then is operated in NOx mode by passing the diluted span gas through the converter. If the converter is working correctly, then the NO reading should recover to its undiluted level as the converter changes the generated NO_2 into NO.

Operation of Smoke Meters

Smoke is the emission of particles that are visible to the naked eye. There are three types of smoke:

1. Black smoke, which is caused by the incomplete combustion of the fuel

2. White smoke, which is caused by the emission of atomized fuel that has not ignited

3. Blue smoke, which is caused by lubricating oil that has passed from the engine and into the exhaust pipe

The level of smoke emission from engines is controlled by legislation in most developed countries. The level of smoke emission from vehicles in the United Kingdom is controlled in three ways:

1. At the vehicle or engine manufacture
2. During the annual vehicle Ministry of Transport (MOT) check
3. By observations or checks performed by the police

Two techniques for the measurement of smoke are in common use. The first technique uses a pump (Figure 7.44) to draw exhaust gas through a filter paper. The blackness of the paper then is given a value by measuring the amount of light that is reflected back from the blackened area. In the second technique, light is passed through the exhaust gas, and the amount of light that is able to reach a detector is measured. In the latter (light extinction technique), light can be passed either across the exhaust pipe (across the duct method) or along the exhaust pipe. The filter paper technique measures only the carbon particles that are deposited on the paper; it cannot detect blue or white smoke. The light extinction technique measures all smoke but cannot determine the color of the smoke. The main advantage of the light extinction technique is that the smoke emission can be measured continuously. With the filter paper technique, only snapshot or steady-state measurements are possible.

When taking exhaust gas samples for analysis, it is recommended that a special probe design (Figure 7.45) is adopted to ensure that the relative gas velocity in the main exhaust pipe does not impinge on the sample rate of the relative analyzer.

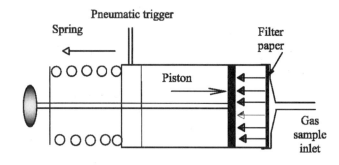

Figure 7.44 Bosch hand pump.

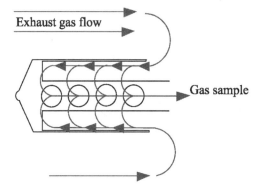

Figure 7.45 Preferred exhaust gas sample probe design.

Operation of Smoke Meters Using the Filter Paper Technique

The simplest smoke meters consist of a calibrated chamber, a spring-loaded piston assembly, a trigger mechanism (pneumatic), and a filter holder. The chamber is evacuated by displacing the piston against the resistance of the spring. The piston is held in the evacuated position by the trigger mechanism. A filter paper is inserted in the holder and clamped using the knurled nut. When the engine is running at the required condition, applying a pneumatic signal to the trigger mechanism triggers a reading. The piston is displaced by the spring, and a calibrated volume of exhaust gas is drawn through the filter paper. The smoke level in the exhaust gas is determined by measuring the light that is reflected from the blackened filter paper.

The following are the main problems encountered with these types of smoke meters:

1. Premature blackening of the filter paper from deposits remaining on the holder
2. Condensation on the filter paper
3. Blocking of the sample probe
4. Heat damage to the sample line
5. Inaccurate calibration of the sample volume
6. Inaccurate calibration of the evaluation unit

The AVL smoke meter shown in Figure 7.46 is, in effect, an automated hand pump in its operation.

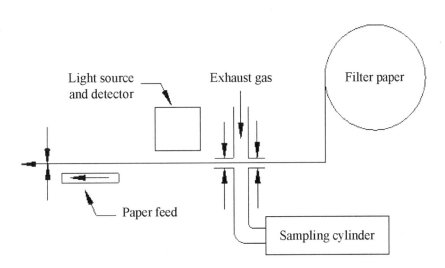

Figure 7.46 AVL smoke meter.

Operation of Smoke Meters Using the Opacity Technique

The smoke meters that use the opacity technique can be divided into two main types:

1. Those that take a sample from the exhaust system (i.e., along the duct) (Figure 7.47)
2. Those that measure across the exhaust pipe (i.e., across the duct) (Figure 7.48)

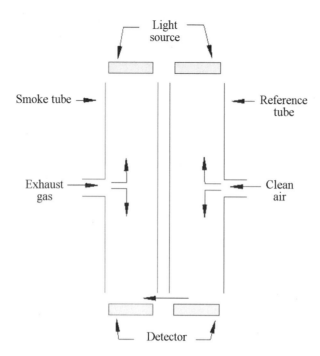

Figure 7.47 Along-the-duct smoke meter.

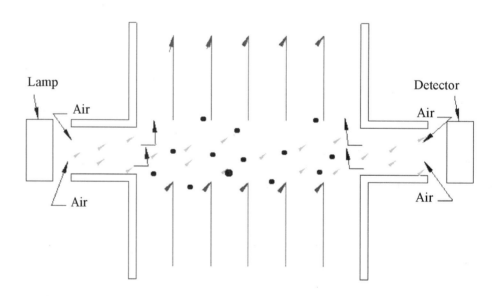

Figure 7.48 Across-the-duct smoke meter.

The amount of light that fails to reach the detector depends on the following:

1. The number of smoke particles
2. The size of the smoke particles
3. The light absorption characteristic of the particles
4. The length of the path of the light beam

For a particular smoke meter, the length of the path of the light beam is constant. Therefore, changes to the number, size, or absorption characteristics of the smoke particles contained in the exhaust gas will alter the amount of light that reaches the detector. The change in the amount of light that reaches the detector will cause a change in the electrical output of the detector. To achieve accurate measurements, it is important to keep the lenses of the light emitter and detector clean. This can be difficult when dirty exhaust gas is being passed continually over the lenses. A curtain of filtered air often is blown across the lenses to prevent the exhaust gas from coming into contact with the lenses and depositing particles on them. If an air curtain is used, care must be taken to ensure that only the minimum amount of air is introduced to the exhaust gas. Otherwise, the exhaust gas will be diluted by the air and will cause an inaccurate measurement. This is especially important at low exhaust gas flow rates.

The following are the main sources of problems with these smoke meters:

1. Contamination of the lenses due to loss of purge air

2. Contamination of the lenses due to water or oil in the purge air

3. Poor optical alignment of the emitter and detector

4. Mechanical or heat damage to the emitter, the detector, and the electrical leads and connections

The main advantage of the across-the-duct smoke meter is the simplicity of the installation. The absence of the need to install an exhaust gas sampling system greatly reduces the chances of errors due to hang-up, condensation, and sample system blockage. Care must be taken to present the exhaust gas to the meter correctly. The meter should not be installed close to any bends or sudden changes in exhaust pipe cross-sectional areas.

One possible disadvantage of across-the-duct smoke meters is that the relatively short length of the path can cause reduced sensitivity to exhaust gas with low smoke content.

Correct Venting of Emission Analyzers

Exhaust gases, traces of ozone, and condensate must be removed from analyzers to prevent a hazardous environment from forming in the analyzer room. The analyzer cabinet should be ventilated to remove any gases that may leak into the cabinet. The exhaust gases and cabinet air should be drawn into a duct that is maintained at a pressure slightly below atmospheric. The waste gases should be discharged into the atmosphere at a location where they cannot be drawn into any buildings or cause any other hazards. Each company and research facility has health and safety data sheets for span gases and ozone.

Factors in Monitoring Exhaust Gases

Some key considerations should be taken into account when monitoring exhaust gas emissions.

The monitoring of exhaust gas emissions must be conducted with great care to ensure the accuracy of results. The following are some key considerations to take into account:

- Temperature ranges
  - Tail pipe—Room temperature up to 300°C
  - Exhaust manifold—Room temperature up to 800°C

- Changes in gas composition
  - Water
  - CO_2 concentration

- High concentration of water
  - Easy to condense

- Particulate matter
  - Nano particulates

- Flow rate changes
  - Several hundred liters per minute to a few thousand liters per minute
  - Transition within one second

- Idle running
  - Strong pulsations in manifold

Compounds Present in Vehicle Exhaust Gases

The following are some of the compounds that are expected to be present in exhaust gases:

- Inorganic
  - CO, CO_2, NO, NO_2, N_2O, NH_3, SO_2, H_2O

- Hydrocarbons
  - Saturated: CH_4, C_2H_6
  - Non-saturated: C_2H_2, C_2H_4, C_3H_3
  - Aromatics: benzene, toluene
  - PCBs: benzpyrene

- Oxygenated hydrocarbons
  - Alcohols: CH_3OH, C_2H_5OH
  - Aldehydes: $HCHO$, CH_3CHO
  - Ketones: $CH_3\text{-}CO\text{-}CH_3$
  - Ethers: $CH_3\text{-}CH_2\text{-}O\text{-}CH_2\text{-}CH_3$

- Particulate matter
  - Soot, soluble organic fraction, sulfides, water

The sources of the compounds listed in the preceding subsection are as follows:

- Intake air compounds: N_2, O_2, CO_2, H_2O…
- Unburned fuel: hydrocarbons, MTBE….
- Complete combustion: CO_2, H_2O
- Incomplete combustion: HCHO, HCs, PM…
- Fuel contaminants: Sulfides…
- Generated from after-treatment: N_2O, NH_3

Emission Testing

When undertaking a total vehicle certified emission test, the vehicle is stored in a standby area that has carefully controlled ambient temperature and humidity. This is to ensure that tests can be repeated at some time in the future with a reasonable degree of confidence in the repeatability. It is necessary to store the vehicle in its soak area for at least 12 hours prior to placing it on the emission rollers. When the test commences, the vehicle is driven following a number of predetermined running cycles that have been designed to replicate the running conditions in each representative country. Figures 7.49 and 7.50 illustrate some of the various test cycles and regulations that are current in Europe and the United States.

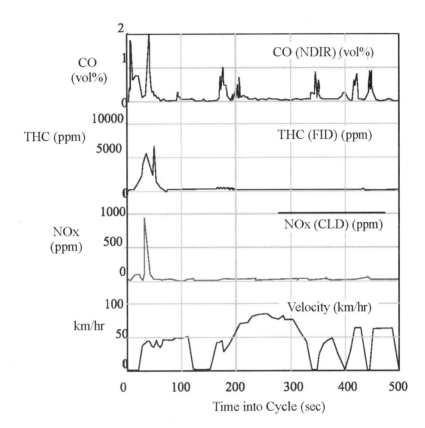

Figure 7.49 Emissions in a driving cycle.

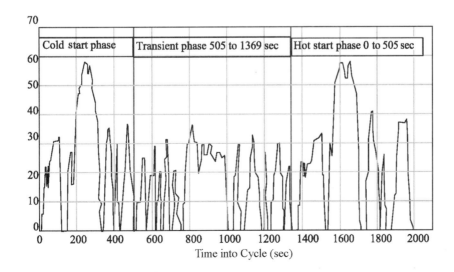

Figure 7.50 U.S. test cycle.

New Exhaust Gas and Particulate Measuring Methods

Several new measuring methods are being advanced. One of particular interest concerns the use of the laser.

Laser

Siemens Process Automation group has developed a high-speed laser device for monitoring the gases in industrial chimneys. This technology is now being applied to good effect in the measurement of NOx, O_2, and ammonia (NH_3) in diesel exhaust systems.

Combustion Fast Response Analyzers for Transient Work

These analyzers have much to commend them, as follows:

- Fast response: full scale in less than 5/1000 sec (5 msec)
- Typical applications: gasoline, diesel, two-stroke, or gaseous fuel engine applications
- Inlet port measurements for the following:
  – Transient EGR calibration
  – Residual mixing studies
  – Spatial EGR distribution
- Exhaust measurements
- Ultra-fast lambda measurement
- Transient fueling calibration
- Catalyst storage investigations
- Cold start studies

Summary

After-Treatment Considerations

After leaving the combustion chamber, the exhaust gases can be converted to safe legislated levels of emissions prior to exiting the tail pipe. This is achieved by oxidizing CO and HC by the addition of oxygen in the presence of heat. NOx will not be reduced by O_2 but requires control either by reducing combustion temperatures or by catalyst control in a reducing (non-oxygen) atmosphere.

The oxidation of CO and HC is enhanced by using a noble metal catalyst, usually platinum or palladium, in the exhaust system. The position of the catalyst is critical to ensure that there is sufficient heat to give early light-off. Lead additives in the fuel render the catalyst ineffective; some oil additives also can cause problems. Excessive over-richness caused by misfire or fueling problems can lead to overheating and burnout of the catalyst formulation (Table 7.5).

Exhaust Gas Recirculation

NOx reduction with exhaust gas recirculation (EGR) is a cost-effective method, but it can have an adverse effect on drivability and fuel consumption. To optimize NOx reduction, the EGR rate normally is increased until the drivability limit is reached, and, in the case of a spark-ignited engine, the spark timing is adjusted to optimum and the air/fuel ratio is adjusted to approximately 14:1. This leads to minimum economy loss with low NOx. The quantity of EGR can be measured by noting the CO_2 in the intake manifold and in the exhaust. EGR is calculated as follows:

TABLE 7.5
NOx REDUCTION/OXIDATION

	Reduction
Desired Reaction	**Undesired Reaction**
$2NO + 2CO > N_2 = 2CO_2$	$CO + H_2O > CO_2 + H_2$
$2NO + 2H_2 > N_2 + H_2O$	$5H_2 + 2NO > 2NH_3 + 2H_2O$
$6NO + 4NH_3 > 5N2_2 + 6H_2O$	$3H_2 + N_2 \leftrightarrow 2NH_3$

	Oxidation
Desired Reaction	**Undesired Reaction**
$2CO + O_2 > CO_2$	$NH_3 + 2O_2 > NO + 3H_2O$

$$\%EGR = \frac{\text{Intake } CO_2 \text{ with EGR} - \text{Intake } CO_2 \text{ without EGR}}{\text{Exhaust } CO_2 \text{ with EGR}}$$

The measurement of CO_2 without EGR is to allow for the background CO_2 and normal internal EGR due to valve overlap. The amount of EGR used will be determined by testing. Generally, as a starting point, look to 8% at low engine loads to 15% at higher values. Reductions of 60 to 70% NOx can be obtained. With increasing EGR, the minimum BSFC point will move to the rich end of the scale. This often is associated with an increase in hydrocarbon emissions (Table 7.6).

Development of Low Emission Fuels

The introduction of low emission levels and new emission-reducing technologies have been linked to the introduction of suitable fuels. Oil companies are working with engine/vehicle manufacturers to develop low emission fuel technology. Some examples of the targets and areas being investigated are as follows:

- Reduced sulfur content for all fuels to allow reliable exhaust after-treatment devices (target level of <5 ppm by 2005). A spin-off from this is reduced H_2SO_4 (acid rain) formation.

- Natural gas (NG) and compressed natural gas (CNG), liquefied petroleum gas (LPG).

- Diesel/water fuel blending for low emissions.

- Hydrogen-based fuels.

- Addition of alcohol to gasoline and diesel (biofuels).

Biofuels and their application are complex problems. Initially, biodiesel has much to offer; however, when further study is undertaken, many problems have arisen, such as the following examples:

- Buildup of ash in the oil from combustion
- Limited shelf life of the base product (six weeks)
- Quality control in a world market and so forth

TABLE 7.6
A GUIDE OF CAUSE AND EFFECT FOR GASOLINE SPARK-IGNITED ENGINES

Engine condition	HC	CO	NOx
Rich mixture	Increase	Increase	Decrease
Weak mixture	Decrease	Decrease	Decrease after 16:1
Ignition advance	Increase	#######	Increase
Ignition retard	Decrease	#######	Decrease
Increased inlet manifold pressure	Decrease	Indirect increase	Increase
Increased exhaust temperature	Decrease	##########	###########
Coolant temperature increase	Decrease	##########	###########
Exhaust backpressure increase	Increase	##########	Decrease
Compression ratio increase	Decrease	##########	Increase
Combustion chamber deposit increase	Increase	##########	Increase initially through to instability
Increased valve overlap	Increase	##########	Decrease
Combustion chamber surface/volume ratio (S/V)	High S/V increases HC	##########	###########
Load	Decrease	Indirect increase	Increase

Chapter 8

Combustion Analysis

Basic Combustion

Before we go any further, let us for a moment consider basic combustion. The *Concise Oxford Dictionary* defines combustion as "the development of light and heat from the chemical combination of a substance with oxygen." We will use the candle as an example. When the candle wick is ignited, we know that combustion is taking place because we can see a flame. If we put our hand over the flame, we can tell at once that it is producing heat. If we place a jar over the candle, cutting off the air supply, the flame goes out. Thus, the flame needs oxygen. What is happening? A general equation for the complete combustion of compounds containing carbon, hydrogen, and oxygen is

$$\text{organic compound (wax)} + O_2 \rightarrow CO_2 + H_2O + \text{heat}$$

The burning of the candle demonstrates this clearly. However, one significant factor has been overlooked: Why is there no noise? Note that the actual process of combustion is silent. In the candle, we have an external combustion device (Figure 8.1). Why then are internal combustion engines noisy? The rapid rise in pressure due to the combustion

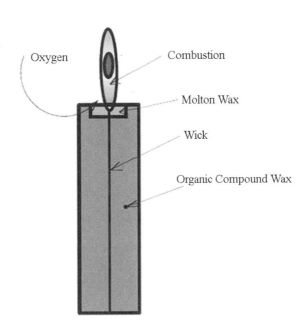

Figure 8.1 *An example of silent external combustion.*

taking place in a confined space gives rise to the noise levels to which we are accustomed. As this chapter progresses, we will investigate this phenomenon further.

If the supply of oxygen is inadequate for complete combustion, the carbon may appear either as carbon monoxide or as the free element, but the hydrogen is always oxidized to water.

Internal Combustion Engine

The internal combustion engine is an energy conversion device that converts thermal energy (heat) into mechanical energy. When hydrocarbon fuel is burned in air, some of the chemical energy contained in the fuel is converted into work. The nitrogen trapped in the cylinder is heated by the energy released when the hydrogen and carbon in the fuel react with the oxygen in the air. A brief step-by-step explanation is given here, using the four-stroke cycle as an example. This also is illustrated in Table 8.1 and Figures 8.2 through 8.7.

TABLE 8.1
FOUR-STROKE CYCLE

Stroke Number	Description
1	The air intake valve is opened, and the piston moves away from the cylinder head, drawing a mixture of air and fuel into the cylinder. (Figure 8.2)
2	The crankshaft rotates and moves the piston toward the cylinder head, compressing the air/fuel mixture.
3	The fuel is ignited at the start of this stroke, when the volume of the mixture is at a minimum. The piston is forced away from the cylinder head by the expanding gas.
4	The exhaust valve is opened, and the piston moves toward the cylinder head.

Intake Stroke

In this stroke, the piston moves down the cylinder, and the pressure drops (negative pressure). The intake valve is opened. Because of the low pressure, the air/fuel mixture is sucked into the cylinder, as shown in Figure 8.2.

Compression Stroke

At bottom dead center (BDC), the cylinder is at its maximum volume, and the intake is closed. Now the piston moves around to top dead center (TDC) and compresses the air/fuel mixtures. The pressure is increased, and the volume is decreased. The necessary work for the compression increases the internal energy of the mixtures—the temperature is increased. Because of the fast compression, only a small part of the energy is transferred to the environment. The compression stroke is shown in Figure 8.3.

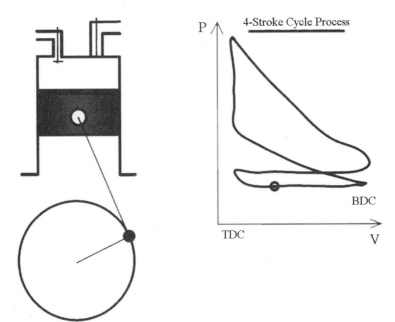

Figure 8.2 Intake stroke.

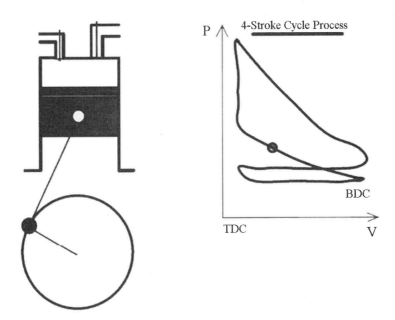

Figure 8.3 Compression stroke.

Ignition Stroke

Near the end of the compression stroke, the ignition starts the combustion, and the mixture burns rapidly. The expanding gas creates high pressure against the top of the piston. The resulting force drives the piston downward. This rapid increase in pressure gives rise to the noise levels associated with combustion. Figure 8.4 shows the ignition stroke.

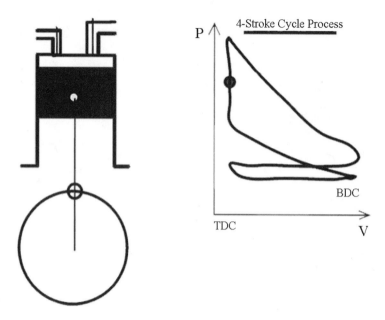

Figure 8.4 Ignition stroke.

Power Stroke

As mentioned, the force drives the piston downward to the crankshaft. (The valves are closed.) The volume is increased, and the pressure is decreased. No more energy is added, and because of this, the internal energy of the gas is decreased, as is the temperature. Figure 8.5 shows the power stroke.

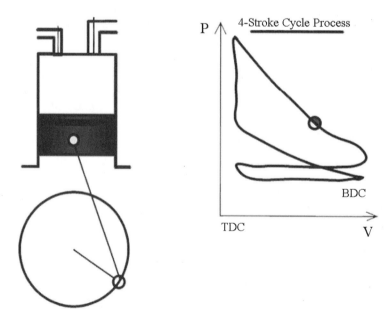

Figure 8.5 Power stroke.

Exhaust Stroke

At BDC, the exhaust valve is opened, and the piston moves up the cylinder. The pressure drops to a level near to atmospheric because of the opened exhaust valve. Exhaust gas leaves the cylinder, and the volume is decreased. The exhaust stroke is shown in Figure 8.6

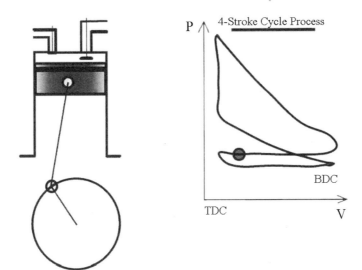

Figure 8.6 *Exhaust stroke.*

When exhaust gases leave the cylinder, the surrounding air cools them. The carbon dioxide gas that is produced will remain as a gas, but the steam that is produced will condense to form water droplets. The intake, compression, ignition, and exhaust strokes are shown in the pressure volume (PV) diagram in Figure 8.7.

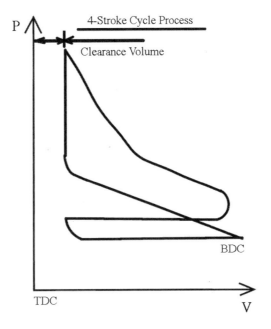

Figure 8.7 *Pressure volume diagram of the four-stroke cycle.*

The pressure volume diagram clearly illustrates the relationship between the pressure in the cylinder and the relative volume in the cylinder between the top face of the piston (i.e., the piston crown) and the cylinder head combustion chamber (i.e., the deck). By studying and analyzing these graphs, one can analyze many critical features of the combustion event, from developed power to pumping losses.

Figure 8.8 shows a most extraordinary graph, produced by Nikolaus August Otto in May 1876. This diagram was taken from the engine by pure mechanical means and bears close resemblance to some up-to-date examples that have the advantage of advanced instrumentation and data acquisition and analysis.

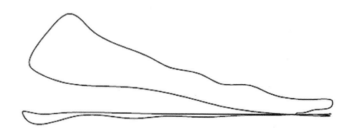

Figure 8.8 A working diagram completed by Nikolaus August Otto in 1876.

Cylinder Pressure Measurements

It is important to note the differences among cylinder pressure versus crank angle, cylinder volume versus crank angle, and cylinder pressure versus cylinder volume because these data form the basis of the classic pressure and volume curves, as shown in Figures 8.9 and 8.10.

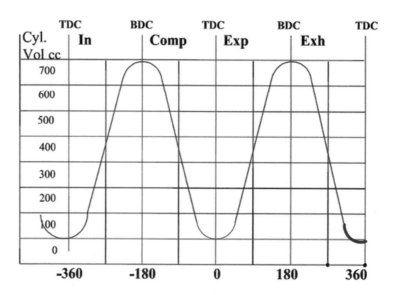

Figure 8.9 Cylinder volume versus crank angle (crankshaft rotation).

Combustion Analysis 167

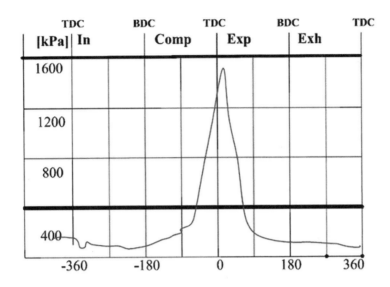

Figure 8.10 Cylinder pressure versus crank angle.

Figure 8.11 shows a typical pressure volume diagram taken from an engine running at 2000 rev/min part load with MBT spark timing. Here we are able to analyze the induction, compression, expansion, and exhaust cycles of a typical internal combustion engine.

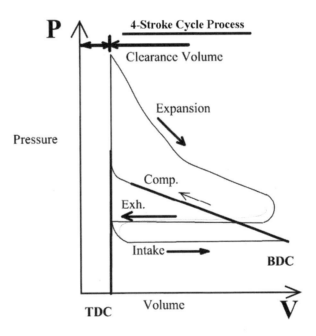

Figure 8.11 A typical pressure volume diagram.

Referring to the Otto cycle diagram already discussed (Figure 8.8), the pressure curve shape will be affected by the following:

- Swept volume/compression ratio
- Valve overlap
- Type of aspiration—atmospheric or charged

On these diagrams, note the intake depression due to the throttle blade in spark ignition (SI) engines (non-direct injection [non-DI]). This is a real power loss because the power to induct the air charge can be considerable.

Efficiency Loss Mechanisms in the Vehicle Drivetrain

Consider for a moment the amount of energy in 2 pints of petrol. This quantity of fuel will enable a vehicle weighing approximately 1 ton to travel 7 to 10 miles. However, if all the energy were released in one instant (i.e., as a controlled explosion, presenting 90% efficiency), then discounting air drag and gravity, the vehicle would be propelled 90 miles through the air. With this in mind, we should and must treat fuel with respect!

Although huge advances have been made in the ongoing development of motor vehicles over the last century or more (Figure 8.12), the automobile remains a very inefficient device, with 100% energy in the fuel tank and only 6% tractive energy at the drive wheels.

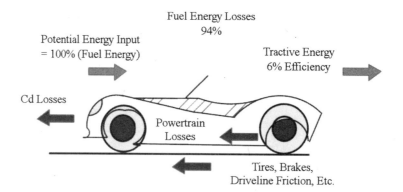

Figure 8.12 Efficiency loss mechanisms in the vehicle.

Losses include the power required to induct and exhaust the air and gas. Figure 8.13 illustrates a summary of the key losses of efficiency.

What Are the Efficiency Loss Mechanisms?

The passage of air, air and gas, and exhaust gases from the inlet to the air filter to the exhaust tailpipe exit point all affect combustion. The objective of engine development is to reduce these loss mechanisms. The diesel engine with no throttle plate has a thermal efficiency of up to 44%. No-throttle DI gasoline (GDI) is seen as a step forward for spark ignition engines. It is evident that there is much room for improvement and indeed an ongoing challenge to the automotive research engineer.

Figure 8.14 shows an ideal or theoretical pressure volume diagram with zero pumping losses; intake and exhaust are a thin line. This is an ideal scenario, with no pumping losses and a complete burn of the mixture.

In reality, due to the design of the internal combustion engine, there are real gas losses, with gas losses past the piston sealing rings, valves not seating correctly, and so forth.

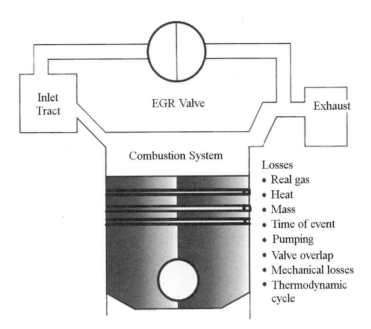

Figure 8.13 Summary of losses.

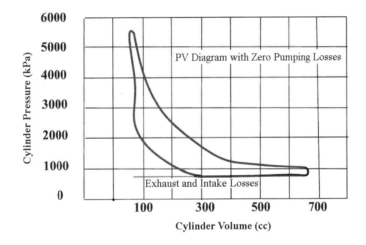

Figure 8.14 Zero pumping losses.

Consider all the gas leakage paths within the engine combustion regime. A primary difference is the heat capacitance of the gas (i.e., the heat is absorbed by the gas). In Figure 8.15, the theoretical maximum cylinder pressure is 5500 kPa, but this is reduced to 4600 kPa due to gas leakage.

Combustion is not instantaneous. It takes time for the flame to propagate and for the pressure due to combustion excitation to rise and apply pressure to the piston mechanism. When the time for combustion to be completed is taken into account, the cylinder pressure drops further to approximately 2800 kPa (Figure 8.16). Actual pumping losses can be considerable (Figure 8.17).

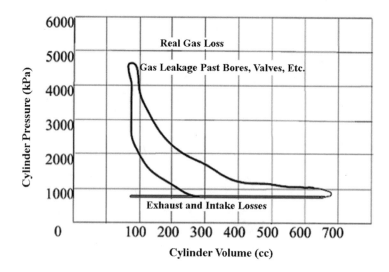

Figure 8.15 Real gas losses.

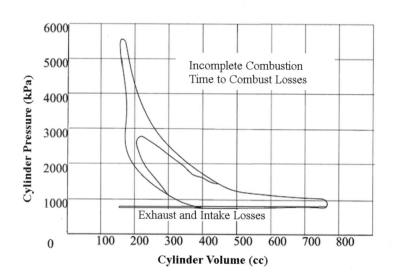

Figure 8.16 Time to combust.

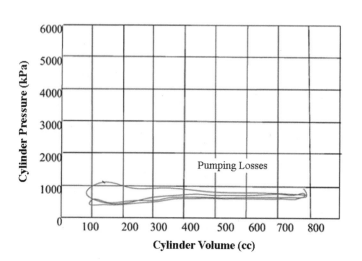

Figure 8.17 Energy lost to pumping.

Heat Losses

Heat is lost from various points in the engine between components. Some of the places and processes where heat loss occurs are as follows:

- Cylinder bore to water jacket
- Combustion chamber deck to water jacket
- Piston crown to connecting rod and then to the oil
- Inlet valve radiated and convection
- Exhaust valve radiated and convection
- Fuel/heat to fuel in evaporation
- Exhaust gases ejected
- Blow-by

Heat loss is a big contributor to an inefficient engine. If heat loss is reduced and the water temperature is increased from 70 to 110°, higher efficiency is achieved.

Unburned hydrocarbons also contribute to inefficiency. Unburned hydrocarbons are likely to occur in crevices, such as at the sides of pistons, by spark plugs or glow plugs, and at valve seat inserts.

At idle in a throttled gasoline engine, cylinder pressures and temperatures are low. The mixture is poor, and there are great efficiency losses. The depression due to the butterfly plate causes uncontrolled turbulence.

The internal combustion engine is an air pump; thus, many efficiency losses occur due to the following:

- Valves not seating
- Blow-by past rings
- Valve overlap
- Friction in rotating the piston crankshaft mechanism

Efficiency losses are illustrated in Figure 8.18, which compares an ideal gas that is approximately 70% efficient to a production gasoline, indirect injection engine with 25 to 28% efficiency.

The summation of all the losses (Figure 8.18) clearly illustrates the potential that lies within the internal combustion engine as we understand it today. Areas in which improvements can be made are as follows:

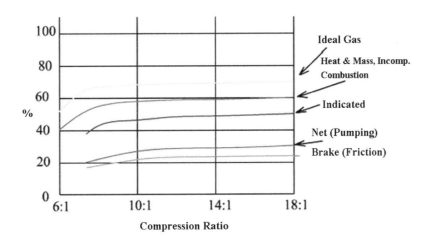

Figure 8.18 Efficiency losses.

- Heat
- Mass
- Time
- Pumping
- Overlap
- Ideal = approximately 65%
- Real = approximately 55%
- Indicated (excludes pumping and friction) = 42%
- Net = 35% (less friction)
- Brake = 28%

Compressed natural gas (CNG) and liquefied petroleum gas (LPG) have lower calorific values than gasoline or diesel and, in the main, produce 12 to 16% less power. When testing an engine, it is apparent that advancing or retarding the crankshaft angle at which the spark is instigated will cause the power to fall off from the ideal MBT (minimum spark for best torque) position (Figure 8.19) where the spark timing is plotted against the average indicated mean effective pressure (IMEP).

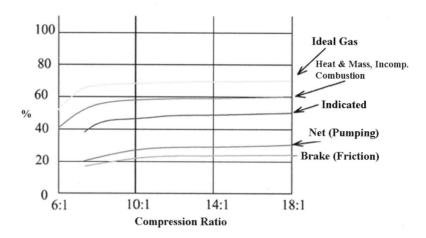

Figure 8.19 Torque versus spark, all cylinders.

If we were to start the burn prior to MBT position relative to TDC, the mixture will burn and build up the pressure holding back the piston as it moves toward TDC, resulting in a loss of power. Conversely, starting the burn after TDC will increase rapidly the volume between the piston crown and the cylinder head deck as the engine rotates, and the buildup in cylinder pressure will be minimal, again resulting in a significant drop in power.

Phasing efficiency losses are a great cause for concern (Figure 8.20). A typical four-cylinder production engine will require different spark timing for each cylinder (injection timing for a compression ignition diesel application). In some instances, these differences can be significant.

It is also a fact that for any given cylinder running at a constant speed and load and at a constant temperature, there will be different performances from cycle to cycle. Figure 8.21 represents a plot of differing spark points, showing the crankshaft angle where 50% of the inducted mass (air/fuel ratio) is burned. The IMEP is noted for five spark angles. Note one measurement per point.

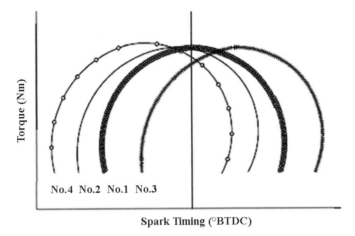

Figure 8.20 Cylinder-to-cylinder phase losses.

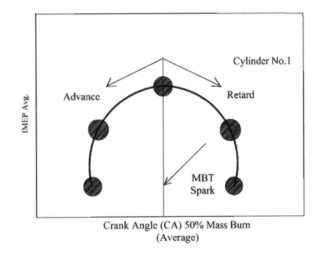

Figure 8.21 IMEP versus CA of 50% mass burned.

When 50 consecutive readings are taken at each selected crank angle, then it will be noted that there is a considerable drift in the IMEP delivered at each selected point (Figure 8.22).

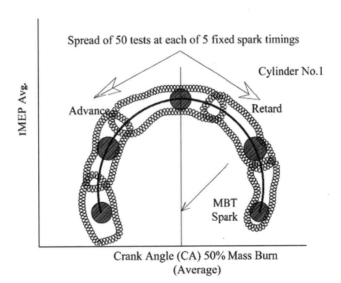

Figure 8.22 Variation in IMEP.

All cylinders on all engines differ from one to another. Taking the measured mean minimum spark for best torque (MBT) as the average is not perfect. To have the maximum efficiency, each cylinder should be optimized individually by measuring the pressure within the cylinder. This is due to cyclic variations, such as the following:

- Spark plug
- Injector signal wire resistance change due to temperature variation
- Valve timing (spring rate variation, natural frequency, and so forth)
- Fuel density (temperature and/or air bubbles)
- Carbon buildup on the injector tip

Electronic control unit (ECU) systems are able to provide differing amounts of ignition advance to individual cylinders at differing engine speed and load conditions.

With the completion of both fuel and ignition loops, both standard development and mapping calibration tools are greatly enhanced with the addition of in-cylinder pressure measurement. Note that in most normally aspirated applications, peak pressure occurs at MBT.

The following is a summary of phasing efficiency loss:

- Overall engine loss occurs when spark timing differs from the overall engine MBT.

- Individual cylinder loss occurs when the individual cylinder MBT differs from the overall engine MBT.

- Individual cycle loss occurs when the individual cycle phasing (CA50) differs from the optimal phasing.

There will always be phasing efficiency loss.

Variability of the spark burn rate always causes a problem in the effort to increase the burn rate and thus the efficiency; however, over-ambitious actions can "put the fire out."

Friction is a huge factor, as is the wind drag created within the crankcase by the rotation of the crankshaft and the movement of the pistons and connecting rods as they push through a fine oil mist (Figure 8.23). Above 8500 rev/min, the power required to push the crank webs and connecting rods through the oil mist is significant.

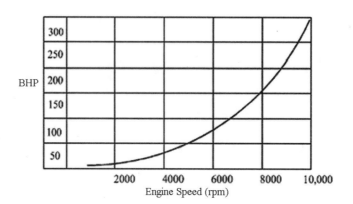

Figure 8.23 Power to overcome rotating and friction losses.

Efficiency Overview

As development engineers and research scientists, we must concern ourselves with the thermodynamic cycle, real gas, heat, mass, incomplete combustion, time, pumping, and phasing. The realistic net thermal efficiency range is 20–40%. Mechanical losses from engine friction and driveline friction and tractive effort efficiency in city driving are only approximately 6%.

Features of Combustion Analysis and Diagnostics

The traditional approach to engine development is to measure the air and fuel entering the engine, the gaseous emissions, the torque output, and the noise. By utilizing the features of combustion analysis, the engineer is able to note what is actually happening within the combustion chamber.

Several types of combustion diagnostics are available, some of which are mentioned here and shown in Figures 8.24 and 8.25:

- Cylinder pressure measurement

- Optical (laser see-through bore and piston single-cylinder special R&D engine) (Figure 8.24)

- Optical (luminosity probes) (Figure 8.25)

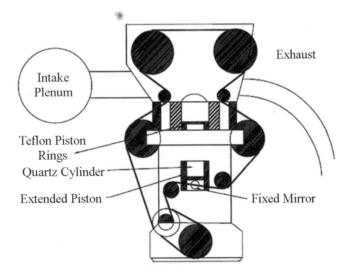

Figure 8.24 Optical engine for laser experiments.

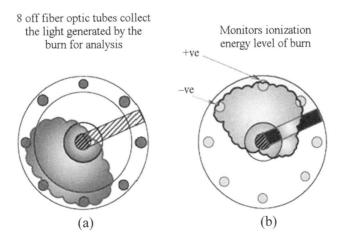

Figure 8.25 Spark plug devices for monitoring flame propagation: (a) AVL spark plug, and (b) FEV ion gap plug.

There are a number of experimental so-called "optical engines" that are manufactured by Ricardo Consulting Engineers in Sussex, U.K., which are typical of many. One is a single-cylinder engine with the provision for changing the compression ratio from 4.5:1 to 25:1. This engine has a quartz bore and an optical plug in the piston crown. A laser beam is directed upward through the transparent piston crown via a hole in the crankcase and a mirror that directs the beam upward through the piston. High-speed video cameras using more than one million frames per minute capture the combustion event though the quartz window in the cylinder liner.

The Austrian instrumentation and research company AVL has developed a spark plug (Figure 8.25(a)) that has a series of fiber optic tubes that monitor the flame propagation for further analysis. The German company FEV has developed a ion gap spark plug (Figure 8.25(b)) that monitors the energy level of the air/fuel mixture burn in its early stages.

Utilizing in-cylinder pressure measurement integrated with crankshaft rotation data, the measurement and interpretation of this combustion pressure is used to determine the following:

- Piston and crankshaft loads

- Torque produced from the burning air/fuel charge equals the IMEP

- Torque required to induct the fresh charge and to exhaust the burned charge equals the pumping mean effective pressure (PMEP)

- Time required for the combustion flame to develop and propagate

- Spark timing relative to MBT

- Presence and magnitude of knock

- Cycle-to-cycle and cylinder-to-cylinder variability

In addition, cylinder pressure combustion analysis can be used to achieve the following:

- Assessment of inlet and/or exhaust port and manifold geometries
- Optimization of the shape of the combustion chamber
- Quantifying compression ratio trade-offs, power/pressure rise/NOx
- Comparison of spark plug parameters
- Selection of valve timing overlap and duration
- Optimization of fuel injector timing and on-time (opening duration)
- Investigation of transient response
- Measurement of mechanical friction
- Automated mapping (MBT, knock, pre-ignition control)
- Calibration optimization

Key combustion performance parameters can be summarized as follows:

- Mean effective pressure
- Combustion phasing
- Cyclic variability
- Heat release

Brake Mean Effective Pressure

The engine output torque at the crankshaft when related to the engine displacement is

$$\text{BMEP} = \frac{\text{Work}}{\text{Volume}} = \frac{\text{Brake work output (Nm) / Cylinder / Mechanical cycle}}{\text{Swept volume per cylinder (liter)}}$$

Note that the brake mean effective pressure (BMEP) is a measure of work output from an engine and not of the pressures in the engine cylinder. The name arises because its unit is that of pressure. Brake mean effective pressure is used to compare the performance of differing engine capacities and the numbers of cylinders. It is expressed in kilopascals (kPa), bar, or pounds per square inch (psi).

$$\text{BMEP}_{\text{bar}} = \frac{1200 \times \text{Power (kW)}}{\text{Engine capacity in liters} \times \text{Engine revolutions per minute}}$$

In considering BMEP, it is necessary to refer again to a pressure volume diagram (Figure 8.26) where we can note that work is a function of pressure and cylinder volume.

Indicated Work—Indicated Mean Effective Pressure

The area enclosed on a pressure volume trace or indicator diagram from an engine is the indicated work done on the gas by the piston (Figure 8.26). The indicated mean effective pressure (IMEP) is a measure of the indicated work output per unit of swept volume, in a form independent of the size and number of cylinders in the engine and engine speed.

Indicated work is made up of the negative work done by the piston due to induction charge compression, plus the positive work done to the piston due to heat release and expansion from combustion, which equals the positive work completed by the cycle, which is the indicated work:

$$\text{IMEP} = \frac{\text{Indicated work output (Nm) per cylinder per mechanical cycle}}{\text{Swept volume per cycle (liter)}}$$

Note that the definition of IMEP used here is not universally adopted. Sometimes (most notably in the United States), the IMEP does not always incorporate the pumping work. This leads to the use of the terms gross IMEP and net IMEP:

$$\text{Gross IMEP} = \text{Net IMEP} + \text{PMEP}$$

where PMEP is the pumping mean effective pressure.

Unfortunately, IMEP does not always mean net IMEP. Thus, it is necessary to check the context to ensure that gross IMEP is not what is intended. The IMEP bears no relationship to the peak pressure in an engine, but it is a characteristic of engine type. The IMEP in naturally aspirated four-stroke spark ignition engines will be smaller than the IMEP of a similar turbocharged engine. This is mainly because the turbocharged engine has greater air density at the start of compression, allowing more air to be burned.

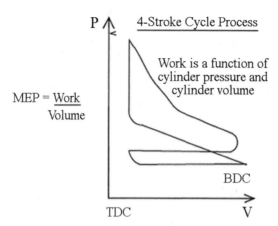

Figure 8.26 Indicated work.

IMEP has been dealt with at length because it is the keystone to further study. We now shall move on to the other family members.

Pumping Mean Effective Pressure

The following equation can be used to determine the pumping mean effective pressure (PMEP):

$$\text{PMEP} = \frac{\text{Swept pumping work (Nm) per cylinder per mechanical cycle}}{\text{Swept volume per cylinder (liter)}}$$

Net Mean Effective Pressure

The following equation can be used to determine the net mean effective pressure (NMEP):

$$\text{NMEP} = \text{IMEP} + \text{PMEP}$$

See Figure 8.27 for a summary of IMEP.

The work output of an engine, as measured by a brake or a dynamometer, is more important than the indicated work output:

$$\text{Brake power} = \text{PbLAN}^1 = \text{Pb}(\text{LAn})\text{N}^* = \text{PbV}_s\text{N}^*$$

where

L = piston stroke (m)

A = piston area (m^2)

n = number of cylinders

V_s = engine swept volume (m^3)

Combustion Analysis

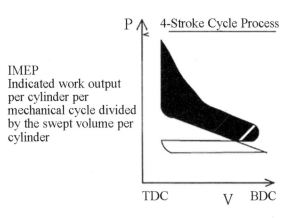

IMEP
Indicated work output per cylinder per mechanical cycle divided by the swept volume per cylinder

Figure 8.27 *IMEP summary.*

N^1 = number of mechanical cycles of operation per second (for all pistons)

$N^* = N^1/n$ = revolutions per second for two-stroke engines, and revolutions per second divided by 2 for four-stroke engines

Frictional Mean Effective Pressure and Mechanical Efficiency

To recap, frictional mean effective pressure is

$$FMEP = IMEP - BMEP$$

Mechanical efficiency is defined as brake power

$$\therefore \frac{BMEP}{\text{Indicated power}} = IMEP$$

The difference between indicated work and brake work is accounted for by friction, and work done in driving essential items (e.g., the lubricating oil pump). Friction mean effective pressure (FMEP) is the difference between the IMEP and the BMEP.

At this juncture, we should consider other key elements that must to be understood.

Indicated Efficiency

When comparing the performance of engines, it sometimes is useful to isolate the mechanical losses. This leads to the use of indicated (arbitrary overall) efficiency as a means of examining the thermodynamic processes in an engine.

Volumetric Efficiency

The volumetric efficiency (ηv) is a measure of the effectiveness of the induction and exhaust processes. However, as some engines inhale a mixture of fuel and air, it is convenient, but arbitrary, to define volumetric efficiency as

$$\eta v = \frac{\text{Mass of air inhaled per cylinder per cycle}}{\text{Mass of air to occupy the swept volume per cylinder at ambient pressure and temperature}}$$

Assuming that air obeys the gas laws, this can be rewritten as

$$\eta v = \frac{\text{Volume of ambient density of air inhaled per cylinder per cycle}}{\text{Cylinder swept volume}}$$

Volumetric efficiency depends on the density of the gases at the end of the induction process; this depends on the temperature and pressure of the charge. There will be pressure drops in the inlet passage and at the inlet valve owing to viscous effects. The charge temperature will be raised by heat transfer from the induction manifold, mixing with residual gases, and heat transfer from the piston, valves, and cylinder. In a spark ignition petrol engine, fuel evaporation can cool the charge by as much as 25°C, and alcohol-based fuels have much greater cooling effects. This improves the volumetric efficiency.

In an ideal process with charge and residuals having the same specific heat capacity and molar mass, the temperature of the residual gases does not affect volumetric efficiency. This is because in the ideal process, induction and exhaust occur at the same constant pressure, and when the two gases mix, the contraction on cooling of the residual gases is balanced exactly by the expansion of the charge. In practice, induction and exhaust processes do not occur at the same pressure.

Types of Combustion Diagnostics

The following discusses the various types of combustion diagnostics including non-heat release and heat release with various algorithms and recommendations.

Non-Heat Release

- IMEP variability—Coefficient of variation (COV) of IMEP
- Combustion limits—Low IMEP, misfire typically indicates an increase in NOx
- Peak rate of pressure rise—Kilopascal (kPa) per degree of crankshaft angle
- Combustion phasing—Location of peak pressure (LPP)

Pumping losses due to the induction of a fresh charge and the exhausting of the burned charge are worthy of study (Figure 8.28). Losses due to the action of sucking air into the cylinder past a near-closed throttle are shown clearly here, with approximately 48 kPa required to draw in the air with a near-closed throttle, and 7 kPa with a near-fully open throttle.

The point of injection in diesel applications and spark relative to crankshaft position also is significant, as shown in Figure 8.29. Changing the point of spark (injection) by 10° crankshaft rotation either side of MBT results in a maximum drop in power of 70 kPa (IMEP) with 10° retard.

Changing the plot to cylinder pressure versus crankshaft angle enables the engineer to not only note the effect of advancing the ignition of on-cylinder pressure, but also to determine the location of peak pressure (Figure 8.30).

Combustion Analysis 181

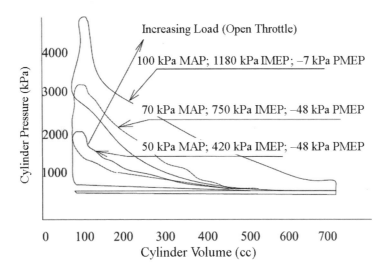

Figure 8.28 Cylinder pressure versus cylinder volume.

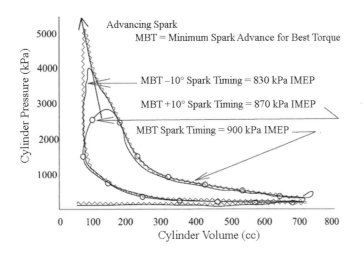

Figure 8.29 Cylinder pressure versus cylinder volume—the influence of spark timing.

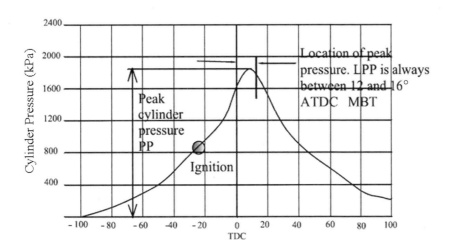

Figure 8.30 Crankshaft angle location of peak pressure.

The determination of peak pressure and the location of peak pressure can be found only by using in-cylinder pressure analysis instrumentation. Note that in most cases, the location of peak pressure is between 12 and 16° crankshaft rotation after TDC when running at MBT.

Heat Release

What is heat release? It is the analysis of cylinder pressure from a firing engine to determine the burn history of the combustion event on a crankshaft-angle-by-crankshaft-angle basis:

- Burn variability—Standard deviation 0–2.5% of mass burn
- Combustion limits—Partial burn and misfire
- Peak burn rate—Percent per degree crankshaft rotation, Joule/degree
- Burn phasing—CA of 50% mass burn

As discussed, it is most dangerous to base development data and indeed mapping on one spot reading (Figure 8.31). Ten consecutive cycles running at the same speed and throttle have differing cylinder pressures and differing locations for the peak pressure point. In this instance, an average of 15° crankshaft rotation was required. Taking and using only one cycle would have resulted in an error of 5° crankshaft rotation and a drop in peak pressure of 50%.

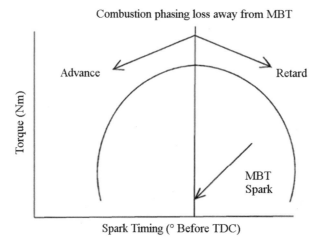

Figure 8.31 Combustion phasing.

Types of Heat Release Algorithms

An accepted approximate method of determining heat release is to use the Rassweiler-Withrow algorithm by taking a standard pressure volume diagram, and then calculate the log of the cylinder pressure and the cylinder volume (Figure 8.32). Note that the compression and expansion lines are parallel to one another.

Assume that the angles of slope for the compression and expansion activities are parallel (Figure 8.33).

Integrating the event between the parallel lines of Figure 8.33 enables us to determine the rate of burn (Figure 8.34). Interpretation of this presents a percentage burn rate per fraction of a degree of crankshaft rotation, thus presenting a significant estimate of heat release.

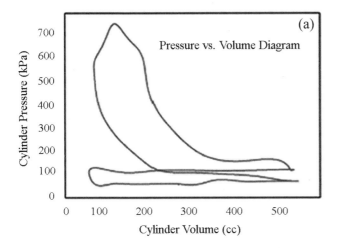

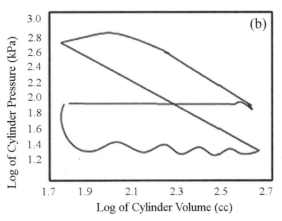

Figure 8.32 Burn rates study. (a) An example of a typical pressure volume diagram. (b) By taking the log of the pressure and the log of the volume, we have this slope.

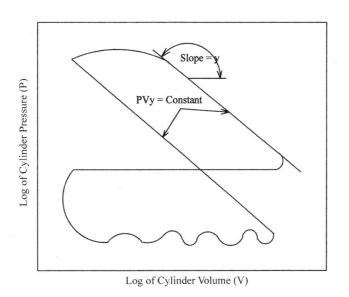

Figure 8.33 Log of pressure versus volume.

The advantages of this method are that it is computationally simple and can be performed in real time. Furthermore, it requires relatively few and readily available inputs. The method assumes that all cycles have 100% combustion efficiency, and that all polytrophic coefficients are equal and constant. The recommended application is any stable operating condition with no partial burns.

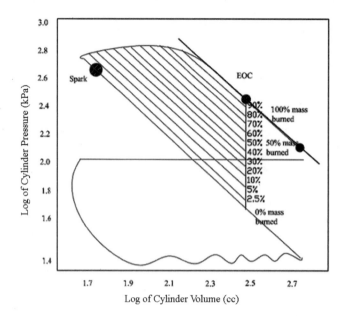

Figure 8.34 Integration of a pressure volume diagram to estimate heat release.

Thermodynamic Heat Release

The First Law of thermodynamic heat release is

$$\delta Q = \Delta U + \Delta KE + \Delta PE + \Delta Q + \delta \omega$$

where

δQ = heat released

ΔU = internal energy = $mC_v \delta T$

ΔPE = potential energy

ΔKE = kinetic energy

ΔQ = heat losses (and mass losses)

$\delta \omega$ = work performed = $\int P dV$

m = trapped mass

C_v = specific heat at a constant volume

T = bulk average cylinder temperature

P = cylinder pressure

V = cylinder volume

Figure 8.35 illustrates the principle of thermodynamic heat release.

One major advantage of this method is that it thermodynamically tracks the mass of fuel burned on the basis of an an individual cycle. This permits the quantifying of several events: partial burns and misfires, residual fraction and residual composition,

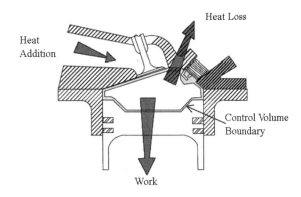

Figure 8.35 Thermodynamic heat release.

and heat losses. Another advantage is that accurate statistics on burn rate variability are provided.

This method assumes that heat transfer can be modeled by an empirical correlation (modified Woschni effect) and that pressure data and other inputs are accurate. Eugen-Georg Woschni was a celebrated physicist who specialized in heat transfer. These other inputs include swirl number, fuel flow, stoichiometry, combustion efficiency, lower heating value of the fuel, combustion chamber surface area, and valve timing.

Recommendations

Combustion evaluation under conditions with high cycle variability can be studied. The cycle-to-cycle variability can be considerable, and in all cases, no two cycles are the same. In Figure 8.36, a standard American power unit, 4.2-liter V8, two-valve cylinder head was run at idle. The crankshaft angle where 2.5% of the fuel air mass was combusted (after TDC) was measured. Note the degree of instability over the first 300 cycles. The probable cause was uncontrolled exhaust gas recirculation (EGR) via valve overlap.

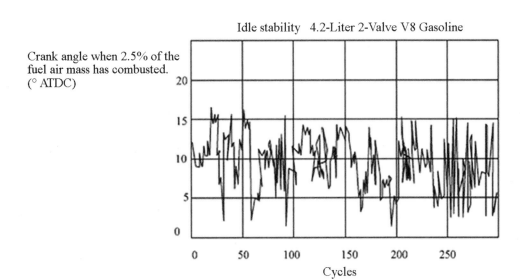

Figure 8.36 Idle stability.

Burn time is critical for a stable idle. A long burn time gives low torque (with low NOx as a trade-off). A short burn time gives high torque.

The potential causes of early burn variability are inadequate ignition (injection) system performance, excessive cyclic air/fuel ratio variability, or excessive dilution (weak charge mixture).

For spark ignition applications, increased spark plug gap gives greater idle cycle-to-cycle stability. For all applications, idle cycle-to-cycle stability is proportional to the degree of valve overlap. High overlap equates with poor idle stability.

Combustion Variability

What is combustion variability? It is the variation in combustion (IMEP) from cycle to cycle and cylinder to cylinder, or in other words, the engine revolutions and developed power vary from cycle to cycle and from cylinder to cylinder. Engine cycles are similar to fingerprints, in that no two are the same.

How Does Combustion Variability Manifest Itself?

The following are some signs of combustion variability:

- Engine roughness
- Cyclic and cylinder-to-cylinder variations in torque and engine speed
- Compromised torque/power
- Lower resistance to knock
- Efficiency losses
- Higher emissions
- Lower fuel economy
- Compromised dilution tolerance
- Compromised spark timing (injection point diesel)

Causes

Combustion variability may be caused by the following:

- Mixture motion at the location and time of spark
- Variation in the amount of air and fuel inducted in each cycle
- Mixing of the fuel and exhaust residuals
- Fuel preparation (droplet size, cone angle, targeting, swirl [barrel and tumble])
- Excessive dilution, exhaust gas recirculation, air, and so forth
- Long burn duration due to poor combustion system hardware design
- Low ignition energy or a small plug gap

Impact

Combustion variability impacts engine performance at all operating conditions. Idle instability typically is driven by variations in fuel flow and exhaust residuals. Part-load variability is driven by fuel flow variations and EGR. Wide open throttle (WOT) combustion instability typically is dictated by variations in airflow.

Note that in the example given in Figure 8.37, there is approximately 38% variation in peak cylinder pressure while running at the same speed/load temperature condition. These readings were taken over a 30-minute period.

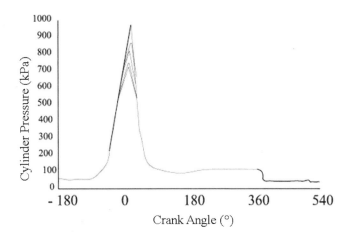

Figure 8.37 Cyclic variation.

Figure 8.38 clearly illustrates an example of variation in burn rate. This example is from a standard production engine and was taken over 10 cycles. Note the relationship between the rate of burn and the mass fraction burned. The burn rate peaked at a mean figure of 18° crankshaft rotation, and the mass fraction burned completed at 38° crankshaft rotation.

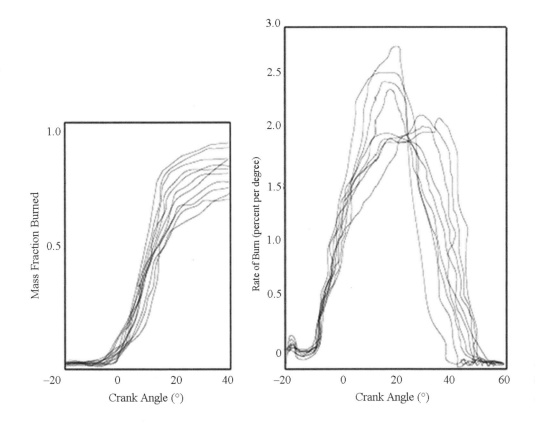

Figure 8.38 Effect of combustion variability on burn rate.

How Is Combustion Variability Quantified?

The most common methods to quantify cycle-to-cycle and cylinder-to-cylinder variability include the standard deviation of IMEP and the standard deviation of revolutions per minute.

- Standard deviation of IMEP quantifies how widely values are dispersed from the mean, that is,

$$\text{STDEV of IMEP} = \sqrt{n \sum_{i=1}^{n} \frac{(\text{IMEP}_i - \text{IMEP})^2}{(n-1)}}$$

where i is the sample of interest and n is the number of samples. (This calculation can be performed on an individual cylinder-to-cylinder basis to quantify cyclic variability or on all cylinders to globally characterize engine stability.

- The coefficient of variation (COV) of IMEP quantifies variability in indicated work by expressing the standard deviation as a percentage of the mean IMEP as

$$\text{COV of IMEP} = \frac{\text{STDEV of IMEP}}{\text{IMEP}} \times 100$$

Although opinions vary, degradation in the drivability of a vehicle can be noticed by the driver and passengers when the COV of IMEP exceeds 3–5%.

- The lowest normalized value (LNV) of IMEP, an indicator of misfires and partial burn cycles, is determined by normalizing the lowest IMEP value in a data set by the mean as

$$\text{LNV of IMEP} = \frac{\text{IMEP min}}{\text{IMEP}} \times 100$$

An LNV of less than 0 indicates a misfire. An LNV of less than 80 indicates a partial burn.

- Standard deviation of engine revolutions per minute (rev/min) due to combustion variability.

- The IMEP imbalance, a measure of cylinder-to-cylinder variation, is quantified by subtracting the average IMEP in the weakest cylinder from the average IMEP in the strongest cylinder, and then normalizing by the mean IMEP as

$$\text{IMEP}_{\text{imbalance}} = \frac{\text{IMEP}_{i,\text{max}} - \text{IMEP}_{i,\text{min}}}{100 \times \text{IMEP}_{\text{engine}}}$$

There are potential pitfalls when taking the average COV of IMEP, and the data can be misleading (Figure 8.39). From Figure 8.39, where four individual cylinders are considered, the engine average COV of the IMEP is 0% for the IMEP at each cylinder and is constant, whereas the IMEP imbalance is 10.5% (i.e., each cylinder has differing IMEP [kilopascal] levels).

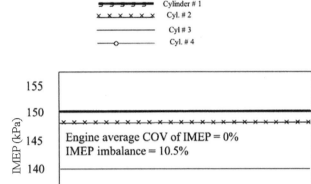

Figure 8.39 Potential pitfall of taking the average COV of the IMEP.

- The root mean square (RMS) of the ΔIMEP (i.e., the highest and lowest IMEP readings) characterizes the difference in work performed in each cylinder event (in the firing order) as

$$\text{RMS of } \Delta\text{IMEP} = \sqrt{\frac{\sum \Delta\text{IMEP}^2}{(n_c \times x) - 1}}$$

where n_c is the number of cylinders, and x is the number of cycles.

Some Thoughts to Ponder

- Do the combustion stability metrics already discussed provide the best measure of combustion stability?
- What does the driver feel?
- What about the difference in work from each cylinder event in firing order?
- Is the phasing of the cylinder events important?

Differential Imbalance Percentage

The differential imbalance percentage (DIP) quantifies the variation in the indicated work done between cylinder firing events by expressing the RMS as a percentage of the mean as

$$\text{DIP} = \frac{\text{RMS of } \Delta\text{IMEP}}{\text{IMEP} \times 100}$$

Referring to Figure 8.40, which illustrates the running of an engine more than 3000 cycles at a constant speed, fixed throttle, and controlled coolant and oil temperature, note that there is a wide-ranging IMEP. The mean IMEP is 126.4 kPa, but the standard deviation of IMEP is 15.4 kPa. The COV of IMEP is 12.6%, and the RMS of ΔIMEP is 24.2 kPa. When reviewing combustion data, no one set of figures will present the full picture.

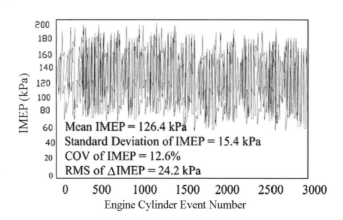

Figure 8.40 Differential imbalance.

Filtering the data can be dangerous if not undertaken intelligently. As illustrated in Figure 8.41, filtering of the data from Figure 8.40 shows a stable engine with the IMEP falling away in a stable fashion after 1000 cycles. These data on their own would indicate an engine where one or more pistons were seizing in their respective bores. That is a totally wrong assumption.

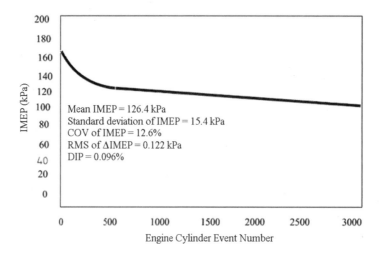

Figure 8.41 Filtering the data from Figure 8.40.

Some Rules of Thumb

Although these generalities do not always hold true, combustion stability usually improves with the following:

- Increased speed and load
- Higher compression ratios
- Lower overlap camshaft timing (valve overlap)
- Higher energy (at the spark plug gap) ignition systems
- Higher temperatures
- Lower humidity

Unfortunately, there typically is a trade-off between high airflow for power and high in-cylinder motion for increased burn rates and less combustion variability, as shown in Figure 8.42.

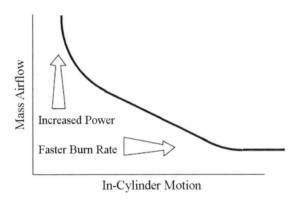

Figure 8.42 Swirl trade-off.

Steps to Improve Stability

The following steps may be taken to improve stability:

- Well-balanced combustion system hardware

- Equal-length, replicated intake and exhaust runners and ports

- Replicated combustion chambers (fast burning)

- Good exhaust gas recirculation (EGR), air, fuel, positive crankcase ventilation (PCV), and purge distribution

- Good fuel injectors

    – Presenting small droplet size

    – With good injector spray targeting (IDI back of valve, minimize wall-wetting)

- Healthy ignition system

- Highest tolerable energy (radio frequency interference [RFI] is a serious concern)

- Largest tolerable spark plug gap (spark plug durability)

- Lowest-resistance high-tension cables (RFI)

- Best compromise camshaft

- Overlap balanced among high-speed power, low-speed torque, and light load combustion stability

- Although a significant factor in high-performance engines, valve overlap can be a problem due to combustion residuals entering the cylinder in an uncontrolled manner (Figures 8.43 and 8.44). When the exhaust and inlet valves are open at the same time, the gas in the exhaust, which is at a higher pressure than in the inlet, will be drawn back into the cylinder. This can lead to idle instability and problems with low-speed throttle or load changes.

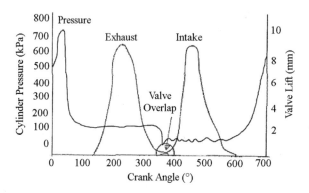

Figure 8.43 Valve overlap.

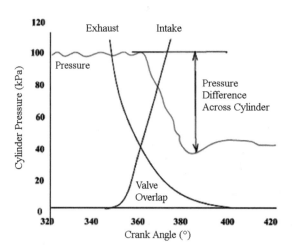

Figure 8.44 Exhaust backflow with valve overlap due to pressure differential induction to exhaust.

Residuals are increased by the following (Figure 8.45):

- Large valve overlap area
- Low speed (more time for backflow)
- Low manifold pressure
- High exhaust backpressure
- Low compression ratio

When applied in a controlled manner, EGR has many advantages. This is known as external EGR. The application of controlled external EGR can increase net thermal efficiency by reducing pumping work.

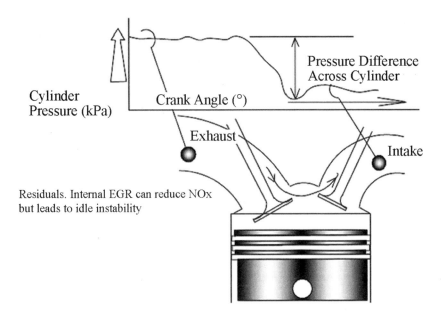

Figure 8.45 Residuals (internal EGR).

How can this be? If the cylinder has been partially filled, then there is less volume for a fresh charge to be drawn inside and thus less power applied in so doing. Exhaust gas recirculation performs the same function as the throttle, without the associated pumping work (i.e., spark ignition only). This has some interesting implications for DI gasoline applications (Figure 8.46). Exhaust gas recirculation also improves fuel economy (Figure 8.47). Furthermore, EGR reduces NOx emissions by reducing the combustion temperatures and the amount of surplus oxygen in the charge with which the disaffected nitrogen molecules can bond (Figure 8.48).

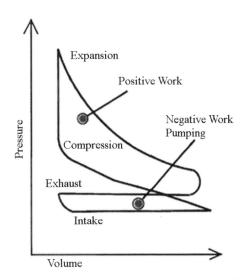

Figure 8.46 Exhaust gas recirculation can increase efficiency.

Too much EGR has distinct disadvantages. A high EGR increases HC emissions, decreases combustion stability, and complicates transient control.

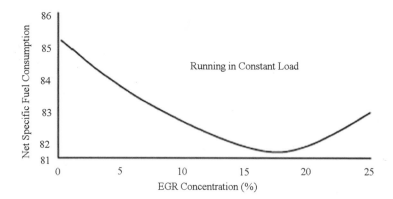

Figure 8.47 Exhaust gas recirculation versus economy.

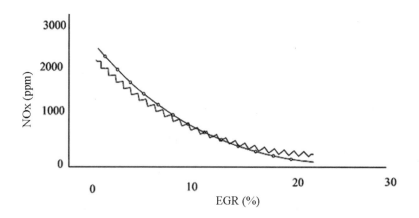

Figure 8.48 NOx reduction with EGR.

Abnormal Combustion

With abnormal combustion, an incomplete burn (misfires and partial burns) can occur. Other problems become apparent as pre-ignition and knock.

Incomplete Combustion

With incomplete combustion, misfires and partial burns occur when flame propagation either is never properly initiated or fails to propagate fully across the combustion chamber prior to the exhaust valve opening. For example, with flame initiation, misfire occurs due to poor spark initiation. Likewise, misfire occurs if the rate of conductive heat losses exceeds the rate of heat production from combustion. Figure 8.49 shows an example of a misfire recorded on a pressure volume diagram.

Using combustion analysis equipment to monitor the IMEP on a cycle-to-cycle basis, irregular misfires can readily be identified (Figure 8.50).

Causes of Misfire

The causes of misfire may be as follows:

- Insufficient ignition energy
- Conditions at the spark plug at the time of spark that are not conducive to ignition:

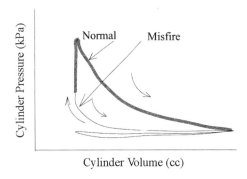

Figure 8.49 A pressure volume diagram indicating misfire.

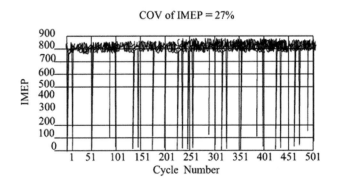

Figure 8.50 Irregular misfires.

- Excessive residuals
- Excessive EGR
- Air/fuel ratio (either too lean or too rich)
- High compression pressures
- Low temperatures
- Mean flow velocity around the spark plug is too high
- Excessive plug fouling

Causes of Partial Burns

The following are causes of partial burns, where the burn is initiated but is extinguished rapidly (Figure 8.51):

- Burn duration is too long
  - Insufficient charge motion
  - Low compression pressures
  - Excessive dilution (residuals, air, EGR)
- Spark timing is too retarded
- Cylinder contents are not well mixed

Irregular misfires due to partial burn also can be examined (Figure 8.52) when noting the IMEP versus the cycle number. Note that on some misfires, the IMEP does not fall to zero but is arrested at approximately 100 kPa.

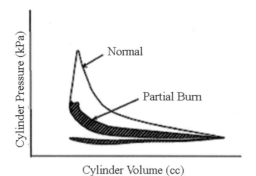

Figure 8.51 Partial burn.

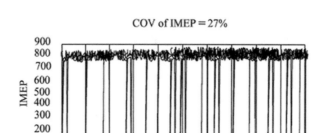

Figure 8.52 Irregular misfire.

Knock

Knock is the explosive spontaneous ignition of fuel/air mixture ahead of the normal propagating flame and the subsequent cylinder pressure oscillations (Figure 8.53).

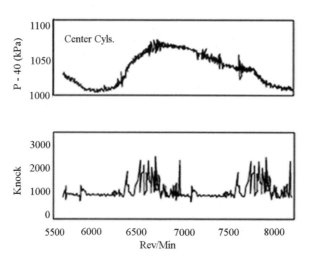

Figure 8.53 Knock.

Knock is not the following:

- **Any combustion-induced noise**—Knock is the result of uncontrolled auto-ignition and will respond to changes in fuel octane. Rumble is the result of high pressure rise rates and will not respond to changes in fuel octane.

- **Detonation**—Typical knock-induced pressure oscillations are acoustic (sonic-noise). Detonation is supersonic.

- **Pre-ignition**—Pre-ignition is the initiation of combustion prior to spark discharge, often the result of a hot spot induced by knock.

Who should worry about knock? The list includes fuel formulation chemists, engine designers, and calibration engineers.

When does knock occur?

- When the engine speed is low and manifold absolute pressure (MAP) is high
- When combustion duration is long
- When temperatures are high (ambient, coolant, combustion chamber surface)
- When the charge dilution is low
- With high carbon deposits
- When the spark advance (or injection timing) is high

Why is knock a problem? If knock occurs, there are ongoing costs to the engine manufacturer, for it is potentially destructive as well as annoying to customers.

What can be done to prevent the onset of knock?

- Improve the quality (octane rating) of the fuel; however, the cost of quality improvement is high.

- Reduce the compression ratio, but at the expense of power and fuel economy.

- Retard the spark, but this reduces torque and fuel economy.

- Enrich the air/fuel ratio, which also increases emissions and fuel consumption.

How should knock be controlled?

- The fuel chemist can do the following:

    - Add expensive blending agents (aromatics and MTBE) to increase octane.

    - Introduce additive packages to minimize combustion deposits on the piston and combustion chamber deck.

- The engine designer can do the following:

    - Design fast-burn combustion chambers.

    - Tighten tolerances to ensure low cyclic variability.

    - Use advanced software to ensure low cylinder-to-cylinder mal-distribution from poorly designed induction and exhaust systems.

    - Ensure excellent structural cooling.

- The calibration engineer can do the following:
  - Introduce fuel enrichment on acceleration.
  - Use spark retard as required.
  - Use trained staff to determine knock by audio means (e.g., trained ear, historic development, engineer's methods).
  - Use an accelerometer (vehicle ECM).
  - Measure the cylinder pressure (modern development engineer), noting the maximum rate of pressure rise, the peak and hold on the filtered pressure trace, and the peak and hold on the smoothed pressure trace.

When studying knock using cylinder pressure measurement equipment, excessive filtering of the input signal can be counterproductive and can hide the onset of critical knock. Care should be taken when selecting filtering options (Figure 8.54).

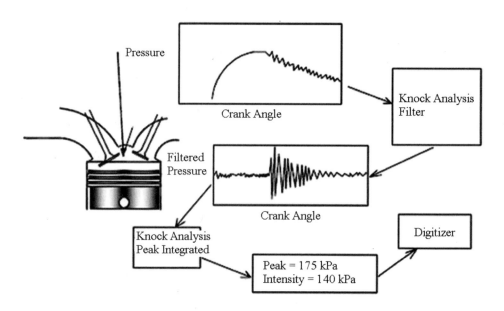

Figure 8.54 Signal filtering.

Figure 8.55 shows an example of a 41-point filter. The basis for filtration is $\dfrac{\sum_{i=-20}^{20} x_i}{41}$, when x_i is smooth.

When presented with a knock problem, the development engineer should ask the following questions:

1. Is knock excessive in all cylinders?
2. Is combustion variability dictating knock?
3. What is the true knock-limited torque?
4. Is the burn rate profile conducive to good knock-limited performance?
5. Can we adequately detect knock?
6. Is the compression ratio too high?

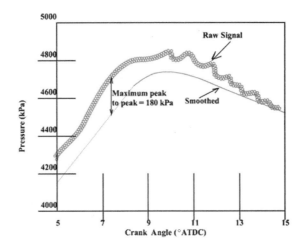

Figure 8.55 Smoothed 41-point filter.

Pre-Ignition

Pre-ignition, or runaway ignition, is ignition in the combustion chamber prior to spark discharge. Referring to Figure 8.56, where will the onset of NOx start? As the pressure within the cylinder rises rapidly, so will the temperature. It will rapidly reach the point where the energy of the burn will disassociate the triple bond nitrogen molecules, and these will bond with the surplus oxygen, thus forming NOx.

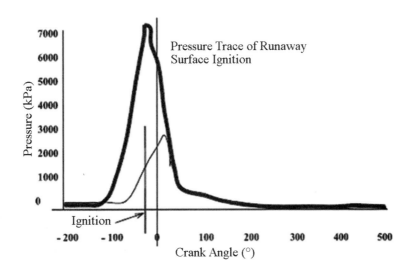

Figure 8.56 Runaway ignition.

Pre-ignition is undesirable because it rapidly produces very high pressures and temperatures in the combustion chamber. It also may cause a piston to melt or break in the middle of the piston crown. (Raw ether used in diesel cold start is a good example of this.) Furthermore, pre-ignition may lead to some other form of catastrophic failure (e.g., crankshaft, connecting rods, valves).

Calibration Issues

It is important to establish a running-on-road footprint of the key speed and load points. Legislated emission points hold priority, of course, as do drivability, performance, power, and fuel economy. Calibration must be an exercise in compromise. For example, when reviewing the required ignition timing for a given throttle and speed load point, we would select the best ignition setting for maximum torque and minimum fuel while considering the EGR concentration (Figure 8.57).

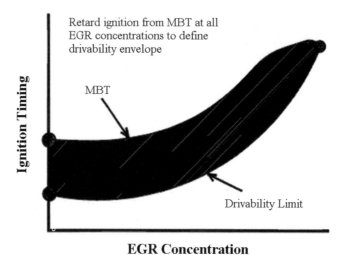

Figure 8.57 Calibration optimization.

The next action would be to superimpose lines of constant NOx (Figure 8.58), where the NOx levels are 400 ppm reducing to 40 ppm. Keeping to the same EGR concentrations and ignition timing, lines of constant hydrocarbons (HC) are superimposed next (Figure 8.59). A similar exercise is undertaken using lines of constant brake specific fuel consumption (BSFC) (Figure 8.60).

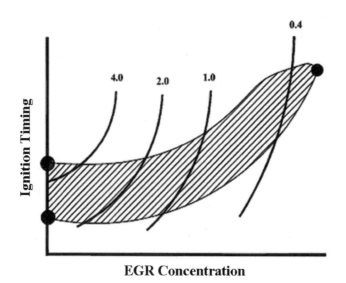

Figure 8.58 NOx contours on a drivability envelope.

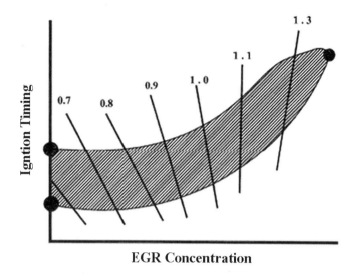

Figure 8.59 Specific HC contours on drivability.

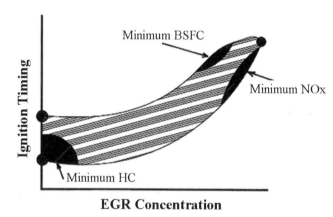

Figure 8.60 Specific fuel of drivability envelope.

It now will become apparent that the condition for best fuel economy is not suited for lowest NOx or for lowest HC. Clearly, a compromise is required. Figure 8.61 shows these optimization conflicts, and Figure 8.62 illustrates a typical calibration compromise.

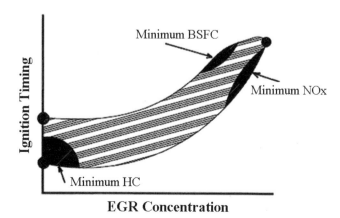

Figure 8.61 Optimization conflicts.

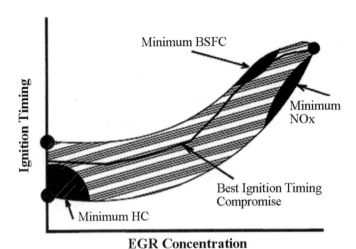

Figure 8.62 Calibration compromise.

Figure 8.63 gives another example of a spark timing strategy. In this case, the point of spark is retarded slightly from MBT to optimize the trade-off between efficiency and gaseous emissions. Retarding the ignition by 1° crankshaft rotation resulted in a 6% drop in NOx concentration.

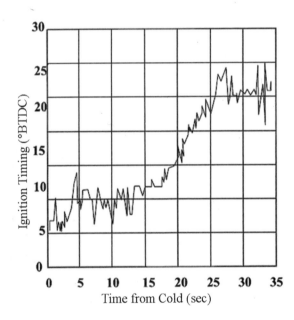

Figure 8.63 Ignition calibration strategy.

Cold Start Calibration

With current emissions legislation, cold start strategy is a most important factor, as can be seen from Figure 8.64, where the majority of HC emissions take place in the first 100 seconds of engine running.

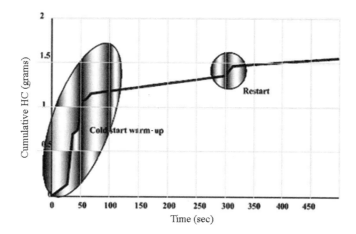

Figure 8.64 Hydrocarbon emissions on start and warm-up.

A typical cold start development program would be as follows:

- Calibrate to a specific combustion stability limit

- Operate at the highest engine speed acceptable from the perspective of noise, vibration, and harshness (NVH) during the cold idle

- Optimize the trade-off between spark retard and air/fuel enleanment to minimize cumulative HC emissions prior to catalyst light-off

Upon cold starting, it is necessary to increase the closed throttle speed (fast idle) to help rapid start-up (Figure 8.65). The faster the engine warms up, the lower the total engine out emissions will be. A special idle air control valve operates following signals from the engine ECU. The difference between the demand and the actual engine revolutions per minute versus time in seconds is illustrated in Figure 8.65. Similarly, ignition timing must change with warm-up (Figure 8.66). The fueling in the first seven seconds of running is critical (Figure 8.67).

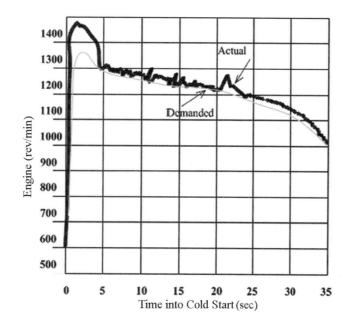

Figure 8.65 Cold start calibration, revolutions per minute versus time in seconds.

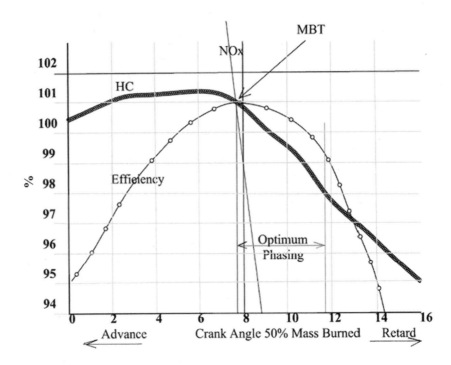

Figure 8.66 Cold start calibration, ignition timing.

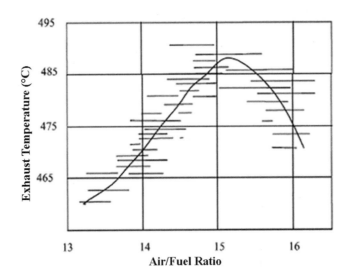

Figure 8.67 Cold start calibration, air/fuel ratio (AFR).

Combustion stability is critical (Figure 8.68) because this affects engine out hydrocarbons (Figure 8.69), and, of course, exhaust heat as the catalyst must obtain a light-off condition as soon as possible (Figure 8.70).

Taking a plot of percentage of the COV of IMEP versus the air/fuel ratio is another tool, in the case of the plot shown in Figure 8.71. Weakening the mixture leads to a higher percentage of the COV of IMEP.

Hydrocarbon emissions (unburned fuel) are affected greatly by the air/fuel ratio (Figure 8.72), and this impinges on the exhaust temperature (Figure 8.73).

Combustion Analysis 205

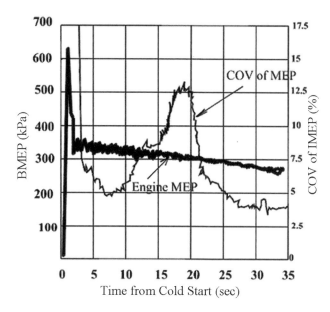

Figure 8.68 Cold start combustion stability.

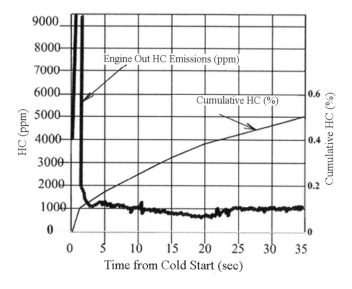

Figure 8.69 Cold start engine out hydrocarbons.

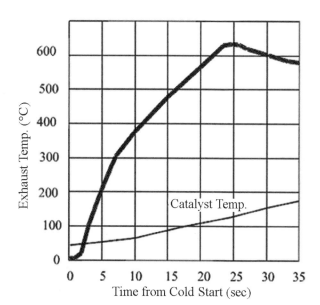

Figure 8.70 Cold start calibration exhaust heat.

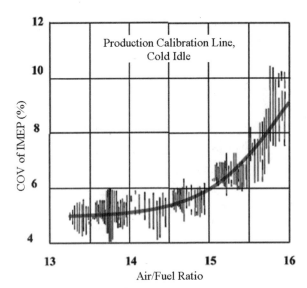

Figure 8.71 Cold start "enlean" sensitivity of COV of IMEP.

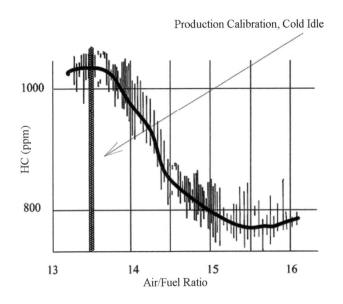

Figure 8.72 Cold start sensitivity (hydrocarbon parts per million [HC ppm]).

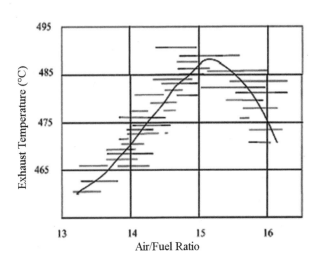

Figure 8.73 Cold start sensitivity (example: temperature).

The overall cold start ignition timing must be cross-referenced with the COV of IMEP (Figure 8.74). Note the production calibration point.

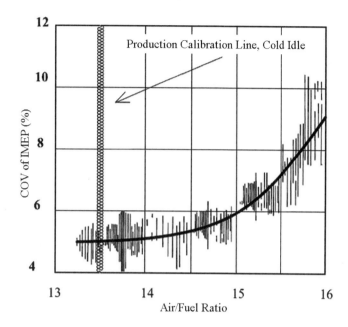

Figure 8.74 Cold start ignition calibration, COV of IMEP.

Most spark-ignited engines are sensitive to ignition retard. Figures 8.75 and 8.76 show the effect of air/fuel ratio and ignition retard on exhaust temperature and hydrocarbons.

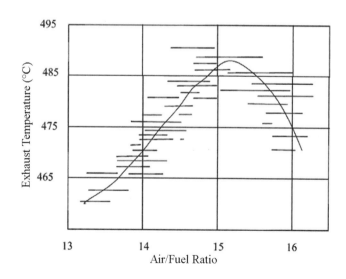

Figure 8.75 Cold start sensitivity to air/fuel ratio versus the exhaust temperature.

Combustion Measurement as a Calibration Tool

- **Combustion phasing**—MAP to an optimum phasing value (CA50 of approximately 10°); use CA50 to check the calibration precision.

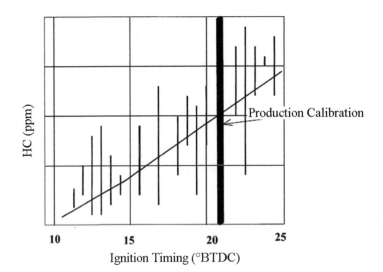

Figure 8.76 Cold start ignition retard sensitivity versus hydrocarbons.

- **Combustion stability**—Manifold absolute pressure (MAP) within acceptable drivability limits; use the COV of IMEP and the IMEP imbalance to check calibration drivability.

- **Knock and pre-ignition monitoring**—Manifold absolute pressure (MAP) within acceptable knock and pre-ignition limits.

- **Onboard diagnostics (OBD) misfire diagnostic tuning**—Tune diagnostic to trigger only on true misfires.

Cylinder Pressure Measurement Hardware

Figure 8.77 illustrates a cylinder pressure measurement system with the following components:

- Calibrated transducer
- High-impedance cable

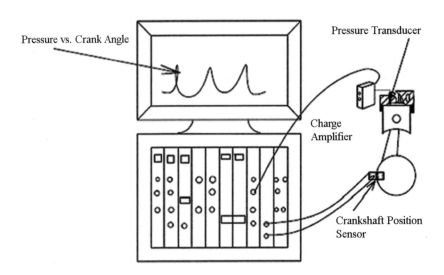

Figure 8.77 Cylinder pressure measurement.

- Charge amplifier
- Encoder
- Digitizer

There are pros and cons for using piezoelectric pressure transducers. The pros include high-temperature and high-pressure capabilities and fast response. The cons are that the transducers are expensive and are good only for fast transient phenomena. Furthermore, the output is not absolute, that is, it must be "pegged." Other potential problems that could arise include sensitivity to both electrical and mechanical noise and to mechanical stress. Poor calibration and moisture in the connections can cause further problems. Carbon accumulation on the transducer diaphragm is problematic, as are inaccurate referencing and thermal shock.

Mounting Considerations

The following are some considerations for pressure transducer installation:

- **Mounting**—Flush mounting avoids resonance issues associated with passages but maximizes thermal shock, especially at the periphery of the combustion chamber (Figure 8.78).

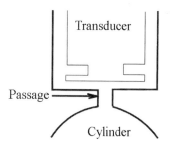

Passage:
1 mm bore
1 to 4 mm long
Minimal clearance between transducer face and inner side of passage 0.10 mm

Figure 8.78 Pressure transducer installation.

- **Torque**—Tighten to recommended torque. Note that the sensitivity will change if the transducer is over-torqued.

- **Leaks**—Check for leaks during operation, and avoid wetting the connectors.

- **Gaskets**—Copper gaskets are recommended. Nickel often is supplied but cannot be recommended.

Referencing Pressure Measurements (Pegging):

When referencing pressure measurements (pegging), keep in mind the following:

1. Transducer output to charge amplifier

2. Charge amplifier outputs voltage to digitizer

3. User must reference voltage to pressure (peg)

$$\text{Pressure} = \text{gain} \times \text{voltage} = \text{bias}$$

$$P = G\varepsilon(\Phi) + \varepsilon_b$$

Is absolute pressure necessary?

Yes	No
Peak pressure	Location of peak pressure (LPP)
Polytropic coefficients	Mean effective pressure (MEP)
Heat release	Coefficient of variation (COV) of MEP

The following are recommended pegging routines:

Procedure	Advantages	Disadvantages
IBDC (set the cylinder pressure at IBDC equal to MAP)	Computationally simple	Accuracy decreases when intake tuning occurs; point measurement is highly susceptible to noise
Average exhaust pressure (set the average cylinder pressure firing exhaust stroke equal to the measured backpressure)	Averaging several measurements minimizes the susceptibility to experimental noise	Computationally intensive; accuracy deteriorates when exhaust tuning occurs
Forced polytropic compression (force the polytropic compression coefficients of each cycle equal to the user input value)	Effective at all operating conditions	Computationally intensive; eliminates the polytropic coefficient as a data quality diagnostic; may be highly susceptible to experimental noise

Recommendations for Using Transducers

The following are recommendations for using transducers:

- Be aware of the limitations of the transducer (i.e., study the manufacturer's specifications; if in doubt, ask the supplier for recommendations in writing)

- Select a transducer with a good thermal shock resistance

- Mount the transducer precisely

- Keep the heat shield and connectors clean

- Use a clean water supply (preferably distilled)

When using high-impedance cable, keep in mind the following:

- The purpose of high-impedance cable is to conduct charge from the transducer to the charge amplifier.

- High-impedance cable is suitable when resistance between the center conductor and the outer shield must be kept high ($10^{13} \Omega$).

- Keep the length of the cable reasonably short.

- Do not allow the cable to sit in liquid.

- Keep the connector clean and dry, and do not use alcohol-based electrical contact cleaners.

- Conduct cable resistance testing.

- The Kistler 5491 insulation tester (price approximately $1,000) is a useful tool.

- Check the cable and transducer/cable assembly.

- If less than $10^{13}\Omega$, then:
  - Clean the connectors with contact cleaner
  - Bake for 2 hours at 125°C

- If the resistance in the cable is still low, check the cable for damage.

The purpose of a charge amplifier is twofold: (1) to convert the charge output from the transducer to a voltage (range capacitor) V: charge/capacitance, and (2) to amplify voltage to produce the desired gain. The following recommendations are offered:

- Use a differential amplifier to reduce the susceptibility to many types of noise.

- Use an amplifier that grounds each engine cycle to help keep data from drifting out of range of the digitizer.

- Choose an amplifier with the capability of offsetting ground to allow more efficient use of the range of the digitizer.

- Select an amplifier that provides information on intra-cycle drift.

Encoders

Encoders are useful for crankshaft position sensing, as follows:

- With an optical encoder, a light gate and optical pickup indicate the crankshaft position.

- Digital magnetic pickup (DI–MAG), 0–5-volt TTL output is dependent on the magnetic field strength and can be used for sensing ring gear teeth.

- With a magnetic encoder, the sine wave output corresponds to the strength of the magnetic field. It is best to use zero crossing rather than some arbitrary amplitude threshold.

- An engine crankshaft sensor generally has insufficient resolution, although 58x systems are adequate for some rudimentary applications (58x = 58 pulses per crankshaft revolution).
  - 360 pulses per revolution for general combustion work are preferred.
  - 3600 pulses per revolution for knock work are preferred.

Figures 8.79 and 8.80 offer details about encoder troubleshooting.

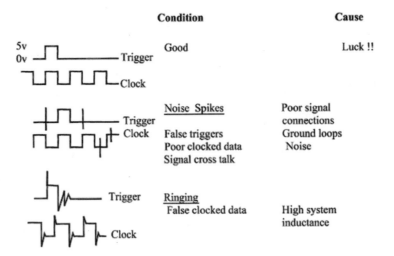

Figure 8.79 Encoder troubleshooting.

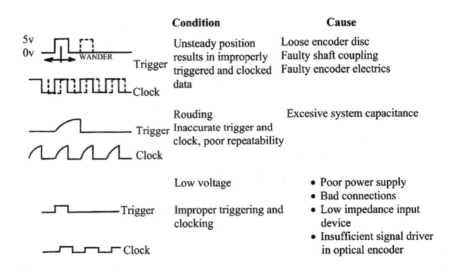

Figure 8.80 More about encoder troubleshooting.

The following are recommendations for using encoders:

- Select an encoder that does not ground to the engine (isolate the case).

- Choose a direct-mount encoder as opposed to one requiring a coupling.

- It may be necessary to photo-isolate encoder signals in high-noise applications. Use a quadrant encoder for applications where the engine may rotate backwards (e.g., cold start tests).

Data Integrity

How is data integrity achieved? It is achieved by understanding the magnitude and causes of variation present in the combustion data acquisition process and then using that knowledge to identify and remove causes that do not occur naturally.

Daily checks will provide the information necessary to understand sources of variability. Daily checks should be performed as follows:

- Record combustion data daily at the same test condition.

- Control all variables to the greatest extent possible.

- Ideally, record data under both firing and motoring conditions.

- Select a firing condition that is representative of the majority of actual test conditions.

- If most of your testing is done at low speeds and low loads, select the daily check condition accordingly (e.g., 1400 rev/min, 15 BMEP psi).

- Perform the motoring test at the same speed and throttle condition (e.g., wide open throttle [WOT]).

- Maintain consistent engine and environmental conditions.

Engine	Environment
Following the same warm-up procedure	Temperatures, inlet air, coolant, oil, fuel pressures, inlet air, fuel
Constant speed	Inlet air humidity
Constant load (brake torque, MEP, MAP)	Same type of fuel
Always conduct motoring and firing tests in the same order	Use the same test technician if possible

Now that you are conducting daily checks and gathering lots of interesting sound data, what are you going to do with it? Plot it on a control chart!

Control Charts

A control chart is a statistical tool used to distinguish naturally occurring variation in a process from variation due to special causes (Figure 8.81). Naturally occurring variation is inherent to any process over time and affects all outcomes. Special causes, or assignable causes, such as a failing pressure transducer or an air leak in an emissions sampling tube, are not always present and do not affect all outcomes.

Control charts are useful in identifying the following:

- Engine problems, such as scuffing pistons, leaking rings, or damaged camshaft lobes

- Insufficient run-in due to the stability of emissions or friction

- Instrumentation problems such as dirty or damaged transducers, failing emission analyzers, or equipment instrumentation drift

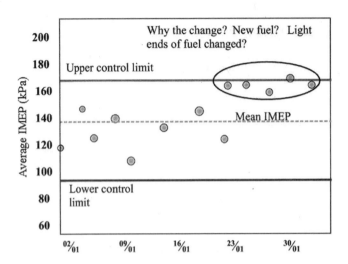

Figure 8.81 Control chart.

Control Chart Set-Up and Maintenance

When setting up and maintaining control charts, it is important to establish control limits as follows:

- Record the data for several days at the daily motoring and firing check points.

- After recording five or six data sets, calculate your control limits (statisticians will want more samples) (19 for 90% confidence in the results).

Note that the assumption has been made that the first five or six data points are in control and are under the influence only of naturally occurring variation.

Furthermore, be diligent and keep good records, as noted:

- When you detect a value that you consider to be out of control (i.e., a wild reading), record that reading in a log.

- Review the charts regularly.

- Recalculate the limits only when a change has been made to the engine/data acquisition system.

- New camshaft, cylinder head, new fuel batch. A change in IMEP could be due to changing a worn camshaft or faulty valve seating for new items, or a new batch of fuel with a higher calorific value (i.e., new fuel versus old fuel stock).

Data to Put into a Control Chart and Why

Firing checks include the following:

- IMEP, PMEP, NMEP, BMEP, FMEP, MAP, rev/min

    - All load indicators should be in agreement.

    - Variations may indicate dirty transducers or recalibration for torque meter.

    - Conclusions should be supported with fuel flow or emissions data.

- HC, NOx, CO, CO_2
- Carbon and oxygen balance
- Air/fuel ratios, air/fuel from oxygen sensors, fuel flow rate, airflow rate, BSFC, NSFC
- Polytropic coefficients, peak pressure (PP), location of peak pressure (LPP), crank angle when 50% of the fuel air mass is combusted (CA50)

Motoring checks should include the following:

- IMEP, PMEP, NMEP, BMEP, FMEP, revolutions per minute (IMEP is a good indicator of the health of a transducer)
- PP, LPP
  - Motoring peak pressures and their location are relatively consistent.
  - PP provides a good transducer check, whereas LPP confirms encoder phasing.
- Polytropic coefficients typically do not vary much, and a little change will cause them to exceed the control limits, so use judgment.

Good Test Practices

Make redundant measures a part of your testing. Typically, any one measurement can be supported by several devices or other measurements, an example being air/fuel ratio. Take the time to understand and apply redundant measures wherever possible. The following are some redundant measures:

- How many ways can you quantify and qualify the measured air/fuel ratio?
  - Carbon and oxygen balance air/fuel ratio
  - Exhaust oxygen sensor
  - Inlet air and fuel mass measurements
  - CO emissions
  - Specific fuel consumption, cylinder pressure, torque
- Good test practices.
- Whenever possible, perform test replications. Do not make a decision based on a single test.
- Randomization.
- Support your data by understanding the variability/uncertainty present in your equipment.
- How many cycles of combustion data should you record?

Some "rules of thumb" to guide you are as follows:

- As variability increases, record more data.

- Idle—Very low speed and light load (World Wide Mapping Points), 500–1000 cycles (1500 rev/min, 15 BMEP psi)

- Part load—Better combustion stability, 300–500 cycles

- High load, high speed, 300 or fewer cycles—Balance the number of cycles against factors such as propensity to knock

- Motoring—Very repeatable pressure traces, less than 300 cycles

Recommended Daily Checks

Accurate chart keeping and the regular use of redundant measures require only a small investment in time to establish and maintain. However, they save many times the investment by reducing development and test time through improved data quality.

Spark ignition engine compression ratio optimization (using in-cylinder pressure measurement as your tool) is a most useful device (Figure 8.82).

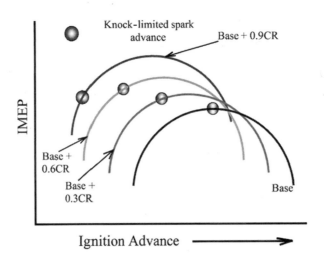

Figure 8.82 Optimize the compression ratio (CR).

The advantages of maximizing the compression ratio are as follows:

- Increased full-load torque through most of the engine speed range
- Reduced full-load combustion-induced engine noise
- Lower peak full-load combustion pressures
- Improved part-load fuel economy (approximately 1.5% per 0.5 ratio)
- Increased dilution tolerance through faster burn
- Improved idle stability via lower residuals

The disadvantages of maximizing the compression ratio are as follows:

- Higher part-load HC and NOx emissions
- Greater reliance on a knock-sensing system
- Higher full-load exhaust temperatures
- Increased likelihood of pre-ignition

Enablers of a high compression ratio are as follows:

- Precise fuel control
- Good cooling of the combustion chamber and the spark plug
- Reliable knock sensing and control methodology
- Low engine-out emissions

Summary

Power and torque from a multi-cylinder engine are dictated by the total contributions of the individual cylinders. Some design features, such as single-point air throttling and/or single-point fuel injection, lead to inherent mal-distribution. Mal-distribution causes the torque curve to be broad and low. Then all cylinders suffer compromised performance.

Trapped mass mal-distribution is the single largest source of mal-distribution in other parameters. However, compensating for mal-distribution in trapped mass by optimizing inlet valve closing reduces the amount of compensation required for other parameters such as spark or injection timing. Achieving maximum individual cylinder performance by reducing mal-distribution substantially increases the overall output of the engine.

Some Calculations Used in Conjunction with Combustion Analysis Work

Combustion Efficiency

The results from the following equations can be used to establish the limit for the new heat release calculations added to the combustion analysis system (CAS). Use the combustion efficiency when the air/fuel ratio is stoichiometric or lean. Use the oxidation efficiency when the air/fuel ratio is rich of stoichiometry. At stable operating conditions when using stoichiometric or lean mixtures, the combustion efficiency will likely exceed 95%. Because the air/fuel ratio is richer than stoichiometric, the combustion efficiency will drop rapidly with an increase in unburned fuel. Therefore, it is difficult to determine the combustion efficiency of 100%. All the fuel could not possibly be consumed because there simply is not enough air present to consume it all. This problem is addressed by the oxidation efficiency that removes the efficiency penalty caused by the energy content of the excess fuel. By looking at the following two equations, it is obvious that they are equal at stoichiometry.

Combustion efficiency (amount of inducted fuel converted to CO_2):

$$\text{Fuel} + \text{oxygen} \rightarrow \text{Heat energy} + \text{water} + \text{carbon dioxide}$$

For the burning of propane, we have

$$C_3H_8 + 5O_2 \rightarrow 3CO_2 + 4H_2O$$

where

C_3H_8 = propane

O_2 = oxygen

CO_2 = carbon dioxide

H_2O = water

Oxidation efficiency (the amount of oxygen converted during combustion):

Oxidation efficiency = Combustion efficiency when the equivalence ratio is less than or equal to 1

Oxidation efficiency = (Combustion efficiency × equivalence ratio) when the equivalence ratio is greater than 1

Equivalence ratio:

$$ER = (\text{Stoichiometric air/fuel ratio})/(\text{Actual air/fuel ratio})$$

Determining the Effective Compression Ratio

This is a reasonable way to determine the effective compression ratio of your engine using measured cylinder pressure. It provides a means of determining the compression characteristics of the engine while comprehending the intake valve closing point, heat loss, and mass losses. The effective compression ratio often will be lower than the measured geometric compression ratio and is affected by manifold tuning. Hence, you probably will find the highest effective compression ratio at the peak torque speed. Engine efficiency and knock will correlate much more closely to the effective compression ratio than the geometric compression ratio. The cylinder pressure data used for this calculation are best recorded on a fully warmed-up engine motoring at wide open throttle.

Ensure that the peak pressure and MAP are in the same units.

Finally, as a recap when using combustion analysis equipment, please remember the following:

- Examine data for signs of measurement error.
    - Accuracy decreases when intake tuning occurs.
    - Point measurement is highly susceptible to noise.
    - The system is computationally intensive.
    - Accuracy deteriorates when exhaust tuning occurs.
    - The system eliminates the polytropic coefficient as a data quality diagnostic.
    - The system may be highly susceptible to experimental noise.
- Set the cylinder pressure at IBDC equal to MAP.
- Set the average cylinder pressure firing exhaust stroke equal to the measured back-pressure.
- Force the polytropic compression coefficients of each cycle equal to user input value.

Advantages are that the system is computationally simple and that averaging several measurements minimizes the susceptibility to experimental noise. Finally, the system is effective at all operating conditions.

Chapter 9

Turbochargers

Introduction

The successful design of a turbocharged diesel engine is highly dependent on the choice of system for delivering exhaust gas energy from the exhaust ports to the turbine, as well as its consequential utilization in the turbine. Almost all the energy of the exhaust gas leaving the cylinder arrives at the turbine. A little is lost due to heat loss, but this loss rarely is more than 5%, unless marine application water-cooled manifolds are used. The design of the exhaust manifold remains critical, as is the valve timing of the engine. A feature that must not be ignored is that the piston must push the combustion products out of the cylinder against a high backpressure imparted by the turbocharger (TC). Chapter 8, the chapter on combustion analysis, details these parasitic pumping losses.

It is the accepted norm that in constant pressure turbocharging, the exhaust ports from all the cylinders are connected to a single exhaust manifold whose volume is large enough to ensure a virtually constant pressure. But technology is changing all the time, and new applications utilizing split gas entries are becoming the vogue.

Pulse Energy

Full advantage is taken of pulse energy. The principle advantage of the pulse over the constant pressure system is that the energy available for conversion to useful work in the turbine is greater. The measurement of exhaust manifold pulse waves relative to crankshaft angle are a prerequisite of turbocharger application and development.

The advantages of constant pressure turbocharging are as follows:

- High turbine efficiency due to steady flow
- Good performance at high load
- A simple exhaust manifold

The disadvantages are as follows:

- Low energy at the turbine
- Poor performance at low speed and low load
- Poor turbocharger acceleration

The following are advantages of pulse turbocharging:

- High energy at turbine
- Good performance at low speed and low load
- Good turbocharger acceleration

The disadvantages of pulse turbocharging are as follows:

- Poor efficiency at high ratings
- A complex exhaust manifold with a large number of cylinders
- Pressure wave reflection problems in some engines

Four-stroke engines are self-aspirating. They have discrete intake and exhaust strokes. Almost regardless of the pressure in the exhaust manifold, piston motion during the exhaust stroke will displace most of the gas. Naturally aspirated engines run with virtually equal inlet and exhaust pressures, and no significant scavenging of the residual gas takes place. When turbocharging, advantage can be taken of the potential difference in manifold pressure to generate a scavenge air throughput to clear the cylinder of residual combustion products. Hence, a pressure drop between the intake and the exhaust is desirable, especially during the period of valve overlap.

Charge Air Cooling

The principal reason for turbocharging is to increase the power output of the engine without increasing the size of the engine. This is achieved by raising the inlet manifold pressure, thereby increasing the mass of fresh air drawn into the cylinders during the intake stroke and allowing more fuel to be burned. However, from the basic laws of thermodynamics, we know that it is impossible to compress air without raising its temperature. As we are trying to raise the density of the air, this temperature rise partly offsets the benefit of increasing the pressure. The design objective is to obtain a pressure rise with a minimum temperature rise because the benefit obtained by raising the inlet manifold pressure is almost halved due to the accompanying temperature rise in the compressor (depending on the efficiency of the compressor). Clearly, it is attractive to cool the air between compression delivery and the intake to the cylinders. A key advantage of charge cooling is that the lower inlet temperature at the cylinders will result in a lower temperature throughout the working process of the engine (for a specific BMEP) and hence reduce the thermal loading. The effectiveness of the intercooler is referred to as the thermal ratio (actual heat transfer/maximum possible heat transfer). Figures 9.1 to 9.6 illustrate the effect of inter-cooling (charge air cooling) on engine performance. (All curves have the same maximum fueling). The lower the air charge temperature, the more dense the air, the greater the oxygen content, and the greater power per stroke capability of the engine. Figures 9.1 and 9.2 illustrate inlet manifold temperature and pressure curves, respectively. For a given volume of air, the lower the temperature, the lower the pressure, and hence the positive effect on filling with a cold charge.

In the example given in Figure 9.3, due to the increased density of the cooled air, there is a 16% increase in the trapped mass within the cylinder due to the intercooler effect.

Figure 9.4 illustrates the efficiency advantages of inter-cooling where the heat transfer to the cylinder walls (energy loss) is greatly reduced when inter-cooling is used. Figure 9.5 presents the fuel consumption advantages of utilizing basic inter-cooling where a fuel savings of 6% is noted at the maximum torque condition. Figure 9.6 illustrates the gains in BMEP that can be realized with inter-cooling. The inter-cooled engine gives more power per cylinder than a non-inter-cooled version with the same fueling, bore, stroke, timing speed, and so forth.

From these figures, it is evident that there are few disadvantages in charge air cooling, and all engine designers and development engineers should endeavor to incorporate some form of charge air cooling inter-cooler, either air-to-air or air-to-water systems.

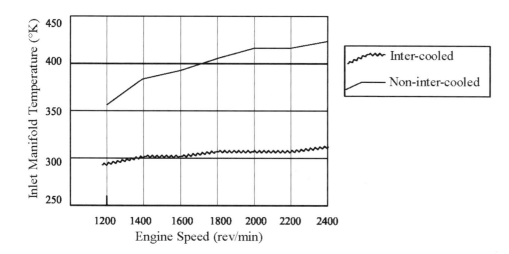

Figure 9.1 Inlet manifold temperature.

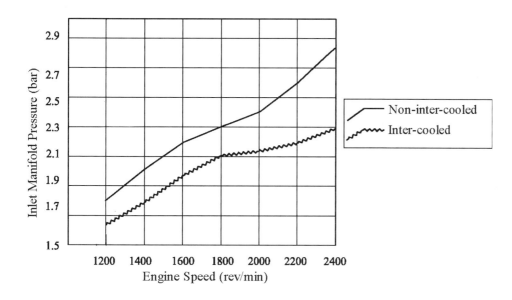

Figure 9.2 Inlet manifold pressure.

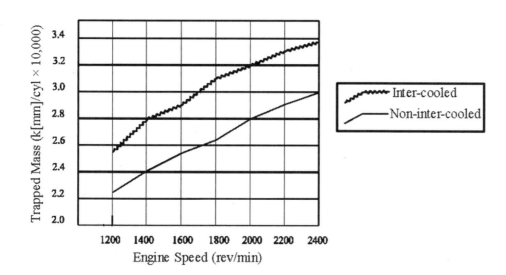

Figure 9.3 Trapped mass in the cylinder.

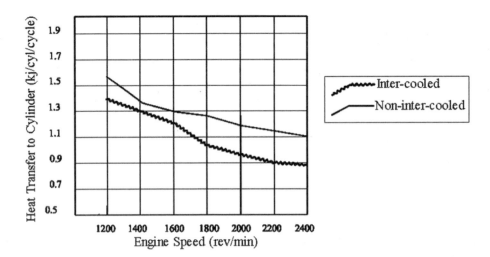

Figure 9.4 Heat transfer to the cylinder.

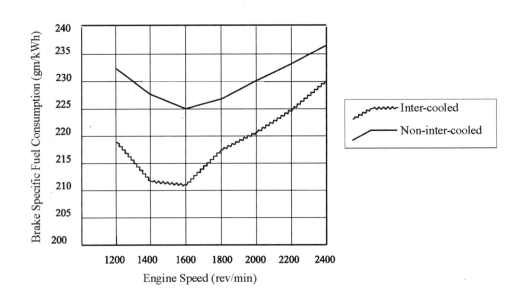

Figure 9.5 Specific fuel consumption.

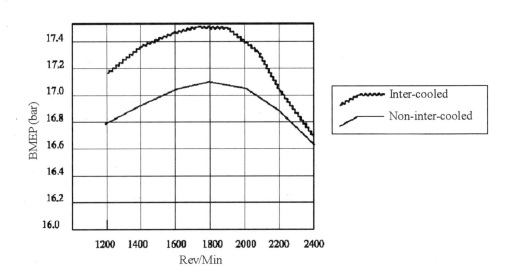

Figure 9.6 BMEP gains with an inter-cooling strategy.

Turbocharger Matching

The performance of a turbocharger is dependent on the angle of gas entry to the impeller, diffuser, and turbine rotor. A correct match will be obtained only when the mass flow rate is correct for a specified rotor speed. It is apparent that the turbocharger will not operate at its high-efficiency flow condition over the complete working range of the engine, and a series of compromises must be made when matching the turbocharger.

The basic size of the turbocharger will be set by the quantity of air required by the engine. This will be a function of swept volume, speed, rating (or boost pressure), density of air in the inlet manifold, volumetric efficiency, and scavenge flow. With a little valve overlap, the volumetric efficiency will be less than 1 and may be estimated from values obtained under normally aspirated (N/A) operating conditions or, indeed, from previous experience. With a large valve overlap, the clearance volume will be scavenged, and some excess air will pass into the exhaust.

The boost pressure will have to be estimated for the engine to produce its target power output, subject to expected thermal and mechanical stress. By assuming a value for the isentropic efficiency of the compressor (total to static), the boost temperature may be estimated. The isentropic efficiency of the compressor can be taken from the manufacturer's compressor maps (Figure 9.7). The final choice of compressor will be made by bearing in mind the complete operating lines of the engine over its whole speed and load range, superimposed on the compressor characteristic map (Figures 9.8 and 9.9).

Figure 9.7 indicates a typical turbocharger manufacturer's performance map, with an engine performance map overlaid. Note that this is a poor map and, if utilized, would result in poor performance at low speed (turbo lag) and a loss of power at wide open throttle (WOT) high speed. Figure 9.8 shows a preferred match with a much higher speed running in a 77% efficient contour loop area.

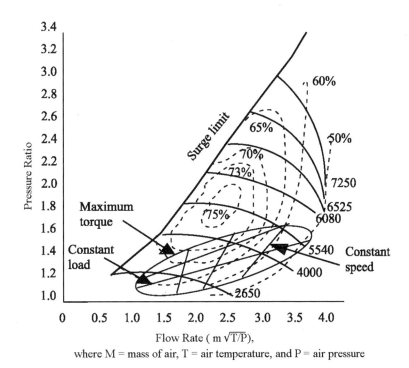

Figure 9.7 Turbocharger is oversized for this map.

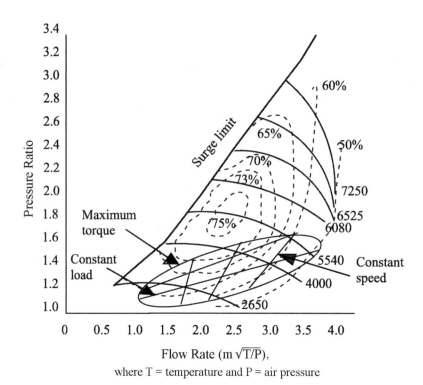

Figure 9.8 Pressure/flow curve.

The type of presentation graph shown in Figure 9.9, known as an oyster curve due to contour lines replicating those of the exterior surface of an oyster shell, enables the development engineer to see at a glance many aspects of the performance of the engine. It also enables the engineer to select a turbocharger with a greater degree of confidence.

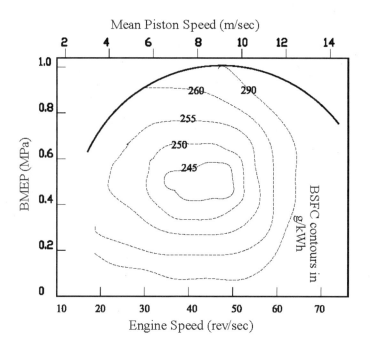

Figure 9.9 Oyster curve for fuel, BMEP, and mean piston speed.

Figure 9.10 illustrates the relationship between boost pressure, compressor discharge temperature, turbo inlet temperature, smoke, and fuel consumption. As is often the case, the development engineer is presented with a compromise. The figure clearly shows that a waste gate or similar device has been introduced to control the maximum boost pressure, in this case holding at 60 kPa.

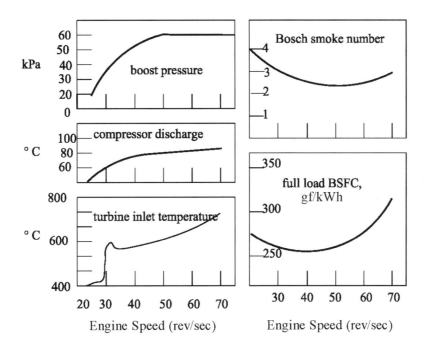

Figure 9.10 Engine revolutions per minute versus boost temperature, smoke, and BSFC.

Matching a turbocharger for an automotive application is difficult due to the wide range of speed and load variations encountered. Although the power required to propel a vehicle rises rapidly with speed, a torque curve that rises as speed falls reduces the number of gearbox ratios and gear changes required (Figure 9.11). Such a torque curve

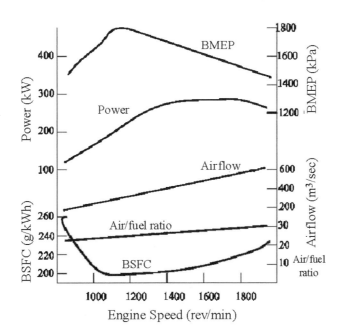

Figure 9.11 A typical power curve, showing the straight line curves of fuel and air.

is said to have good torque backup. Pulse turbocharging is essential for good torque backup at low speed.

The torque curve in Figure 9.12 shows limits to BMEP, set by the allowable smoke limits, maximum designed cylinder pressure, exhaust temperature, and turbocharger compressor revolutions per minute. As always, the development engineer is presented with a compromise situation.

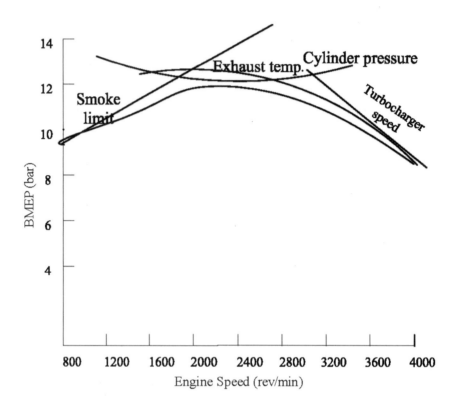

Figure 9.12 Compromise limits.

Development of advanced common rail systems and associated equipment has introduced additional freedom to vary fuel delivery over the speed range of an engine. The turbocharger matching process must be linked closely with the fuel system matching, even after optimum injection rates, pressures, nozzle sizes, and swirl have been achieved.

As we have seen, the main role of the turbocharger is to increase the mass airflow rate into the engine by compressing the induced air. The performance is affected by the area-to-radius ratio of the turbine, the final trim of the compressor, the turbine wheels, and the wheel radius. In the case of waste gate or air box blow-off applications, the major problem lies in the fact that there is a significant performance trade-off between high-end and low-end speed performances and transient running conditions. Efforts to improve the gas exchange at the low-end (low-speed) regions will impact the overall volumetric efficiency of the device because increased backpressure occurs at high speeds, with loss of performance and, of course, reduced compressor efficiency.

In the past, the engineer had few options other than using a range of turbochargers of differing sizes and trying to find some cost-effective means of isolating them as required. An elegant solution has been the development of the variable vane geometry turbocharger (VGT), which enables the utilization of the advantages of both a small and a large turbocharger in the same casting (Figure 9.13).

Figure 9.13 An example of a variable vane geometry turbocharger (VGT) mechanism.

Figure 9.14 shows a comparison of boost pressure between a variable vane geometry turbocharger and a conventional waste gate turbocharger at full load. From this comparison, we can see that the boost pressure can be increased throughout the whole engine speed range, and it is effective in improving power output by the simple expedient of supplying more air to the engine. In addition to boosting the power density of an engine, a variable vane geometry turbocharger has the potential to improve efficiency by reducing the pumping loss. By optimizing boost pressure at part load, where pumping losses have an important part in fuel economy, significant improvement in economy can be achieved.

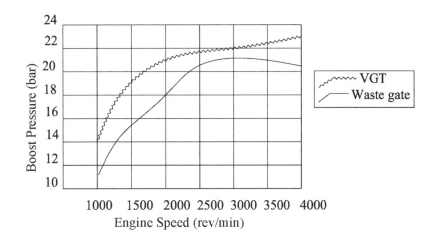

Figure 9.14 Comparison between a variable vane geometry turbocharger (VGT) and a waste gate turbocharger.

Figures 9.15 through 9.17 show the effects of VGT applications on BMEP, BSFC, pumping IMEP, and smoke against engine speeds at full load.

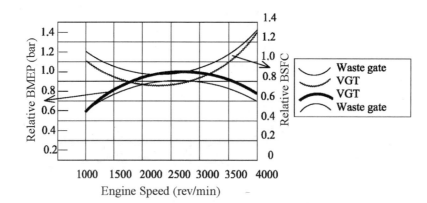

Figure 9.15 Effect of variable vane geometry turbocharger applications on BMEP.

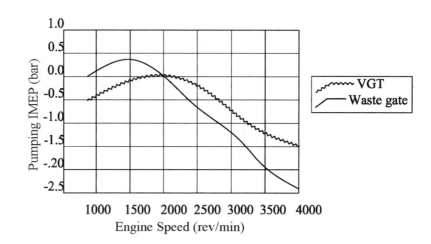

Figure 9.16 Effect of variable vane geometry turbocharger applications on pumping IMEP.

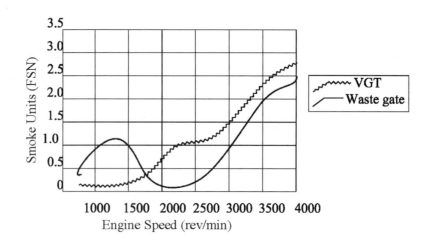

Figure 9.17 Effect of variable vane geometry turbocharger applications on smoke.

The selection of the desired pressure ratio is a key consideration when selecting a turbocharger. Over-boosting can damage the engine. The base compression ratio and valve timing all must be considered. When large engines are reviewed, the strength of the crankcase can be a limiting factor in determining boost pressure ratios (Figure 9.18).

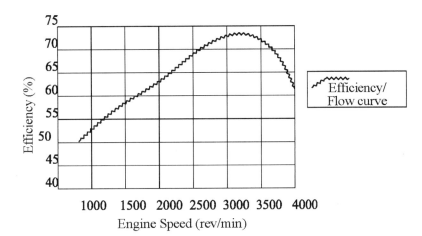

Figure 9.18 *Pressure ratio of 2 efficiency versus turbine flow.*

Chapter 10

Fitting Operations Within the Test Cell Best Practices

In the engine testing industry, great care, vigilance, and diligence are prerequisites. Methodical, meticulous, and, in some cases, well-documented records of fitting tasks are required to assure high-quality service. When considering this statement, bear in mind the following points. Many of the engines and much of the test equipment used in the test industry are extremely sensitive and expensive and require careful handling. Most damage occurs in the installation and removal of engines into and out of the test cell areas.

Many of the engines used in testing are prototype units that cost huge amounts of money to develop to the stage where they can be tested in a test cell, and spare parts will be at a premium in terms of cost and availability. Those engines that are not prototype units may be test engines, developed purely for the test function, with limited access to spare parts. Test programs cost a great deal of money, time, and effort to set up and run; therefore, a lack of care in fitting work at any stage may invalidate the test results. Assured quality is everything. The test industry is built on this cornerstone of quality workmanship and reliability. With these points in mind, here are some recommended rules for fitting in the test cell area.

Stripping

All parts, fasteners, and other removed items should be collected and placed in a container or storage area away from the bed plate and engine assembly. All components that are removed should be laid out in the order in which they were removed. Note that it is important to count all fasteners and associated washers, nuts, pins, and so forth. If anything is dropped, find it before you go any further, in case it is forgotten later. Unless you can account for any missing items, you are gambling with a large amount of money, time, and effort if things go wrong, and things have a habit of doing just that.

Checking and Inspecting

All parts should be checked for serviceability as they are removed. Those components found to be unserviceable should be labeled correctly, in accordance with company procedures, and stored separately from the reusable components. Certain key components bear closer examination and, in some cases, replacement as a pre-emptive measure, rather than risk a failure on test.

Rebuilding

Replace all parts with new seals and gaskets as required and assemble in a logical and methodical manner, counting all the fixings back onto the engine assembly. The component layout and fixing containers should be empty upon completion of the rebuilding, apart from items that have been renewed (e.g., washers, lock nuts). Any items that have been replaced with new ones must be removed from the test cell to reduce the risk of accidental refitting or unaccountability later, that is, having these "spare" items lying around at a later fitting operation may cause confusion, particularly when counting fixings back onto the engine during rebuilding.

Golden Rule for Handing Over

If, for any reason, you must pass the fitting operation to another technician (e.g., shift changing), the golden rule is to ensure that "what you fit is secure." If you do not have the time to secure it, then do not fit it. The next technician then can carry on work, knowing that what has been fitted to date is securely fitted. This will save time in rechecking that everything is fastened properly.

Cleaning

After fitting, clean the engine, bedplate, and cell floor thoroughly. This will help you to quickly spot any leakages that develop after the fitting operations are complete and the engine is running. The earlier a leak is found and remedied, the less likely damage and wear will take place internally in the engine, which could invalidate the test results. There also will be less likelihood of interrupting an expensive test for the sake of remedial attention to components that have failed. Experience with engines will give the test technician the wisdom to double check, or even renew, suspect items prior to a test run as a matter of course, thus reducing the risk of failure later. Furthermore, the test cell will be a safer place in which to work if it is clean. If the client wishes to visit the engine test cell, that client will find it clean and safe, giving a good impression that you indeed are looking after his or her project. Technicians should note all leaks and component failures in the appropriate test documentation (or computer file), because all information recorded during the test may prove important and save time and cost later. Likewise, that information will always add value to a client's project report.

Quality

Quality is everything in engine testing, and good workmanship is one of the main contributors to a high-quality company. Best practices are always open to amendment for the sake of improvement; however, they should never be ignored.

Commonly Occurring Incidents When Testing Engines on Dynamometers

Damaged Valve or Piston and Bore

This type of incident could lead to total engine failure. If this occurred, replacement of the whole unit would be required because small items lost into the inlet tract may not be recovered prior to engine start-up. This may be due to ignorance of the loss of the item(s) in the first place, which is why such items should be counted as a matter of course. The "gambler" who cannot confirm the final resting place of lost or dropped items is a liability and should not be trusted with high-profile or expensive tests, if trusted at all. One engine failure can result in costs to the company in figures much higher than the salary of that test technician and possibly several of his or her colleagues.

Timing Belt "Jumping a Tooth or Three"

This situation could happen if a small item jammed on the crankshaft pulley, lifting the belt and upsetting the crankshaft-to-camshaft timing (valve timing). Although this may not cause serious damage to the major assemblies in the engine, the interruption of the test may prove extremely costly, particularly if the test must return to the start of a long stage. In addition, the engine may have to be stripped completely to establish its integrity before proceeding with the test.

Sheared or Snapped Prop Shaft

This problem could occur due to a lock-up on a universal joint (U/J), possibly caused by a small item entering the area of that universal joint. The shearing of a high-speed prop shaft can have devastatingly dangerous implications for a test technician in the cell, not to mention the damage to the dynamometer and associated equipment.

Summary

Technicians who have failed to follow the golden rules before handing over fitting operations to other technicians (e.g., on shift hand-over) have allowed faults to be built into the test engine by not carrying out a fitting task to its completion, including security and adjustment of components fitted. Although this technician may have initiated the fault process, the following shift technician may have been guilty of presuming all was in order. However, we are discussing basic principles of quality workmanship here, and it is not considered excessive to expect good standards from colleagues, particularly in a test environment.

Failure to correctly identify and label substituted components results in wasted test time. Future tests may be run with previously tested items that have been wrongly selected and fitted, perhaps due to incorrect or insufficient identification and labeling. Remember here also that the client may wish to examine any failed components and to evaluate them in the light of your incident record logs. If you have not accurately identified those components and tied them in some way to the incident log, then under subsequent evaluation your testing operations will be viewed as suspect in accuracy, diligence, and responsiveness to the client's needs.

Remember that a good test technician possesses the following attributes:

- Provides quality workmanship
- Is diligent
- Demonstrates consistency
- Has an inquisitive mind
- Is interested in self-development
- Is knowledgeable
- Possesses experience, which will come with time
- Loves engines, which is a bonus

Chapter 11

The Basic Internal Combustion Engine

Introduction

When considering the performance of the internal combustion engine under test, the technician must have a strong appreciation of how the engine functions with regard to its individual systems.

The systems referred to here are shown in Figure 11.1 and are as follows:

1. Engine lubrication system
2. Engine cooling system
3. Engine induction system
4. Engine exhaust system

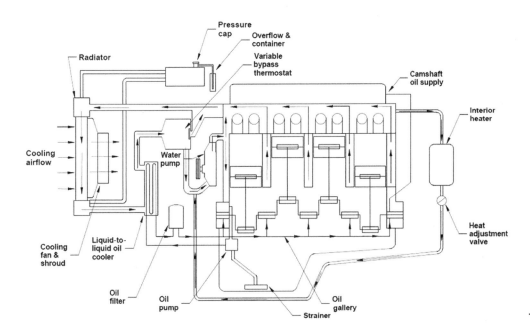

Figure 11.1 Basic engine fluid flow paths.

Obviously, there are other areas for consideration within engine systems (e.g., fuel); however, the preceding systems are those that are monitored most frequently over a wide range of tests. This section intends to provide a refresher on the purpose, operation, and instrumentation of the individual systems for test purposes, together with some advice about typical faults frequently found on test. Also, it is worth mentioning here that the technician must consider that each of the individual systems reacts with other systems within the engine, as well as having its own specific purpose. This factor can be a great help to the test technician in diagnosing faults on test. For example, where a suspiciously high coolant temperature may initiate alarm warnings on the test control equipment, the technician would do well to pay attention to the lubricant temperature, as a way of confirming the overall levels of the engine temperature. The problem may simply lie in improperly sited or faulty coolant temperature sensors.

A thorough understanding of the engine systems and their relationships with each other are absolute musts for the competent test technician. Note that the test data will always tell the knowledgeable observer what is happening inside the engine. The key is in being able to interpret the data and make use of them. A test technician should always question the data being churned out by the test control equipment. Knowledge in hindsight is of little use to the client. (It may help the technician, provided he or she learns from the experience, but that may be at the cost of an important client.) A short refresher in engine technology is offered here but only on a basic level of engine construction.

Engine Lubrication System

The primary purpose of the lubrication system is to supply all mechanical moving parts with a barrier/film of oil on which they can slide. Without oil, the engine will seize due to the effects of friction and the heat generated by the friction, causing the components to weld themselves together in extreme cases. If total seizure does not occur, then the engine could (at least) be severely damaged by the effects of overheating to the point at which the lubricity of the oil is destroyed.

The secondary function of the lubrication is to assist in the removal of heat from within the engine. The effect of reducing the temperature of the engine oil is achieved when the oil in the sump is cooled by the airflow over the sump as the vehicle is driven along the road, or by the introduction of oil coolers into the standard lubrication system.

Lubrication System Components

Several components make up the lubrication system of an engine, as follows:

- **Sump**—Stores a sufficient quantity of oil and aids cooling by its location in the airflow path of the vehicle

- **Strainer**—Primary filtration that removes large contaminants from the oil prior to the oil reaching the oil pump via the pickup pipe

- **Pickup pipe**—A means of getting the oil from the sump to the oil pump

- **Oil pump**—Pressurizes the oil and maintains a steady delivery to the engine components

- **Pressure relief valve**—Controls the pressure of the oil delivered to the engine components

- **Oil filter**—Removes fine particle contaminants from the oil
- **Oil filter bypass valve**—Allows unfiltered oil to reach the engine components if the oil filter becomes blocked
- **Main oil gallery**—Provides the oil distribution route for the oil to reach various areas within the engine assembly

Sump

The sump has two main areas:

1. Bulk oil-holding pan area
2. Sloping raised area that allows oil to drain back to the bulk holding area

The sump can be constructed from pressed steel or cast aluminum. Within the bulk oil-holding area, there will be vertical baffle plates. These are fitted to prevent oil from flowing (e.g., under acceleration or braking) from one end of the sump to the other, thereby causing starvation of the oil pump as the opening of the pickup pipe is uncovered. Horizontal baffle plates also are fitted above the normal oil level to prevent crankshaft rotational windage from creating a vortex hole in the oil held in the bulk oil area of the sump. Again, this would cause the engine to be starved of oil if the baffle were not in place.

Strainer

The strainer filters the oil to remove large contaminants before the oil enters the oil pump via the pickup pipe.

Pickup Pipe

The pickup pipe is flanged and is bolted on one end to the oil pump casing. The other end of the pickup pipe is located in the bulk oil area of the sump just above the base and below the normal level of the oil. Its location in the bulk holding area means that the pipe is immersed in the deepest amount of oil. Fitted to this lower end of the pickup pipe is strainer gauze, which filters out the large particles of contaminants.

Oil Pump

There is more than one type of oil pump design available from which to choose, such as the vane, gear, or eccentric gear types. The oil pump either is located in the sump area or can be mounted to the side of the engine block. The engine block is cast with two gallery oil-ways from the oil pump mounting position. One allows oil to be drawn from the sump by the oil pump. The other allows the pressurized oil from the oil pump to be fed onward to the oil filter housing.

Pressure Relief Valve

This valve normally is located within the oil pump casting and consists of a known rate-of-compression spring and a two-part sliding valve. As the oil pressure supply from the oil pump increases, the spring pressure is overcome, allowing the inner valve slide to move off its seat. This limits the oil pressure supplied to the engine.

Oil Filter

Where possible and unless the test specifications state otherwise or if filter products are being tested, the engine manufacturer's own or recommended product should always be fitted for test purposes. Two main types are in use: (1) the paper element type and (2) the canister type.

- **Paper element**—This type has an outer casing, which has a through-bolt securing it to the engine block at a point where the oil-way is exposed, thereby allowing oil flow into the filter body. Inside this casing is a disposable paper element. The whole filter assembly also can be situated away from the engine, provided there is sufficient flow through its connecting pipe-work so that the performance of the engine lubrication system is not affected.

  Note that an important point here when servicing this type of oil filter is that with modern oils and their additives, the effects of the spent oil on humans are not fully known but are believed to be highly carcinogenic and must be handled carefully.

- **Canister**—This type is primarily the same as the paper element type but is contained in a sealed casing, with a thread provided to allow attachment to the cylinder block. From a test technician's point of view, this type is not as easily examined as the paper cartridge type in the event of engine oil contamination (i.e., cutting open the element to look for foreign bodies). Some canister oil filters are messy to change, and the previous comment about careful handling is equally relevant with the canister type.

Oil Filter Bypass Valve

Within the oil filter is fitted a bypass valve that allows the oil to be supplied to the engine when the oil filter element becomes blocked. Blockage will occur when service intervals are not kept up to date. Then the oil supplied to the engine will not be filtered, which means that engine wear will increase if the oil filter is not changed. The internal bypass system is fitted to both paper element and canister types of filters, but the fact that the bypass valve is in operation is not checkable by visually inspecting the engine or the exterior of the oil filter.

Main Oil Gallery

The main oil gallery provides the distribution route that allows the oil to reach various areas of the engine assembly, including the following:

- Main bearings
- Big ends of the connecting rods (via gallery drillings in the crankshaft)
- Small ends of the connecting rods (via internal rod drillings)
- Piston crown and skirts (via drilling outlets through the connecting rods)
- Camshaft bearings

Oil also is supplied to the cylinder head via drillings in the cylinder block, passing through the cylinder head gasket to aligned drillings in the cylinder head, and normally fitted with a restricting device. If the engine is of the overhead valve type (i.e., with the camshaft situated in the engine block), then the oil feed to the cylinder head supplies oil to the valve rocker shaft assembly. All the oil used within the engine is recycled to the engine sump via drain-back galleries cast into the block and cylinder head. This drain-back is by means of gravity; consequently, the entire inner surface of the engine will be covered with oil.

Instrumentation of the Lubrication System

Within the lubrication system, the information to be obtained is as follows:

- **Oil temperature**—Sump and gallery. The temperature of the oil is measured using thermocouples.

- **Oil pressure**—Pump and gallery. The pressure of the oil is measured using gauges, or, if a more accurate measurement is required, the system can be linked by pressure pipes to a transducer.

- **Oil flow rates**—Pump, across the oil filter, main oil gallery, and feed to the cylinder head. The flow of the oil in the system is measured using flow turbines that are fitted into the system pipe-work. The positioning of the different types of instrumentation normally is as close to the engine as possible. In some cases, the instrumentation must be fitted into the engine systems. To facilitate this, the engine must be partially or wholly stripped, with modifications, drilling, and tapping of components and so forth carried out, and the engine rebuilt to ensure that any devices so fitted are secure and will not affect the engine or its performance in any way.

Health and safety considerations—All oils, used and unused, should be treated as a health risk; therefore, personal protection should be used (e.g., protective gloves, overalls). In addition, the disposal of oils and waste is an environmental issue and should be conducted in the correct manner. Companies should have a policy statement, usually contained within the quality manual, regarding the safe disposal of waste products from the testing operation, and it is always recommended that the technician follow these.

Engine Cooling System

The primary purpose of the engine cooling system is to remove the heat absorbed by the engine assembly as a result of the combustion process, as well as friction heat generated within the engine assembly as components move against each other at high speeds.

Cooling System Components

The following components are part of the engine cooling system:

- Engine block, with integral water jacket
- Cylinder head, with integral water jacket
- Bypass system
- Water pump
- Thermostat
- Thermostat housing
- Radiator
- Expansion tank
- Pressure relief cap
- Cooling fan
- Pipe-work
- Coolant

Engine Block

When an engine is to be liquid cooled, the designer ensures that the engine block has sufficient and efficient flow path for the cooling liquid. This allows coolant to circulate around the areas of the engine where combustion heat is found, which is mainly in the region of the cylinder liners and the combustion chamber in the cylinder head assembly. This flow path is called the water jacket and normally is cast into the engine block in the production process. As stated, the main considerations are that the coolant must surround the cylinder liners to absorb heat that is conducted through the cylinder walls from within the cylinders. There must be provision for a coolant pump to be housed within the flow path or water jacket. At the top surface of the block, where the cylinder head is to be mounted, there must be sufficient paths to allow the coolant to pass to, and to return from, the cylinder head.

Cylinder Head

The main function of the coolant within the cylinder head is to remove heat from around the cylinder head combustion chamber area. A position for the engine thermostat also may be included in the design of the cylinder head.

Bypass System

The bypass system allows rapid warm-up of the coolant to operating temperature. While the thermostat is closed, the bypass circuit (some manufacturers utilize the heater matrix for this circuit) allows coolant to circulate continually around the engine block and the cylinder head, thereby retaining the heat from the engine within the engine assembly until the engine has reached the optimum temperature for the most efficient running.

Water Pump

The water pump is used to pump water for purposes such as engine cooling, to promote high volumetric efficiency, to ensure proper combustion, and to ensure mechanical operation and reliability.

Thermostat

The thermostat acts as a controller for the engine coolant temperature. Therefore, the opening temperature set point of the thermostat that is fitted is determined by the most efficient operating temperature of the engine. The thermostat valve can remain closed until a set temperature is reached (i.e., the amount of heat in the engine needed to maintain its required efficient running temperature). Once this temperature is attained, opening the thermostat valve allows heat to be carried away to the cooling system, preventing overheating due to retention of excess heat energy.

In a vehicle, the thermostat helps to get the engine warmed to its optimum heat range quickly because it is manufactured to be in a normally closed state. However, in engine testing, the thermostat usually is fitted but is set in a normally open state by jacking it open. This is necessary to retain the flow-restricting characteristics of the thermostat within the engine cooling system, while eliminating the interference of the thermostat in the engine temperature as test conditions are altered during the test procedure.

The choice of thermostat, with regard to the ambient temperature at which the engine is to be operated, is the manufacturer's consideration. If the ambient temperature were low, then a high-temperature-opening thermostat would be fitted so that the engine operates as near to optimum temperature as possible, and vice versa for high-ambient-temperature areas. Test technicians may need to be aware of this if testing an engine for use in extreme-temperature environments.

Thermostat Housing

This is a housing in which the thermostat will be held, and it can be positioned to the front top of the cylinder head or located remotely from the head, as long as it is situated within the flow of the main water flow around the engine. The housing normally is connected to the upper section of the radiator.

Radiator

The radiator usually is located to the front of the vehicle within the direct flow of air. It may hold most of the coolant; the remainder is held in the engine and associated pipe-work. The construction of the radiator is such that it has two tanks connected by small-diameter pipes. These pipes are spaced apart, and the resultant gaps are fitted with corrugated cooling vanes. The normal layout is one tank at the top of the radiator and the other at the bottom. These tanks are connected to rubber hoses, which in turn are fitted to the thermostat housing (top tank) and the coolant pump (lower tank). In engine test cells, the radiator seldom is used because the cooling and temperature controlling of the engine under test conditions are managed more effectively by the use of heat exchangers and a raw water supply, with temperature and flow rate controlled more closely.

Expansion Tank

This is a separate tank that is fitted to the system and serves three purposes. First, as the engine warms up, the coolant expands. This means that extra room within the system is required to accommodate this expansion, hence the title "expansion tank." The second purpose of the expansion tank is to allow any aeration of the coolant to separate out within the expansion tank. Thus, to allow both functions, the tank is never filled to its maximum but only to its halfway point. Third, the vapor pressure suppresses boiling of the coolant in the system. The expansion tank often is used in test-cell engine cooling to allow both functions to be retained.

Pressure Relief Cap

This simple device serves two purposes. First, it allows the engine to run at higher temperatures without boiling by pressurizing the coolant system. Second, it relieves excess pressure buildup within the cooling system. The cap is made so that the cooling system operating pressure is maintained to the cap blow-off pressure. The range of temperatures at which pressure relief caps are operated depends on the engine manufacturer's requirements. Thus, great care must be exercised because an incorrect pressure cap can affect the results of tests conducted on the engine, which will alter the characteristics of the cooling system—in some cases, quite dramatically.

Cooling Fan

This fan draws air across the cooling surfaces of the radiator. Because tests seldom are run with radiators fitted to the test bed, the cooling fans normally do not remain fitted to the test engine.

Pipe-Work

Pipe-work around the engine can be made of steel, aluminum, or rubber and is used to link various components of the system to the engine. Rubber pipe serves best because it acts as an isolator, as the engine is rubber mounted to the bedplate and the heat exchangers are solid mounted. Any engine vibrations will be absorbed by the rubber hoses and will not be transmitted to the heat exchangers, which would cause failure.

Coolant

Engines require protection against both corrosion and, in low-ambient-temperature climate testing, freezing. Modern antifreeze solutions also improve coolant efficiency in high temperature ranges. Therefore, the engine coolant is made up of water and antifreeze. The main component of antifreeze is ethylene glycol, and this reduces the risk of freezing, depending on the mixture strength (commonly 50% water and 50% ethylene glycol). At this strength, the boiling point of the coolant mixture rises to 109°C. Ethylene glycol also has a higher boiling point than water, so it allows engines to be run at higher operating temperatures without the risk of boiling. Of primary concern to the technician is that with so many variants available, the correct product and solution strength must be chosen for the right engine or engine test requirement.

Fire warning—Care must be taken because ethylene glycol has an ignition point of 125°C; therefore, the mixture strength must be kept to the absolute minimum required with a small safety margin. Because of this, the most common mixture strength used is 50-50 water/ethylene glycol. If a sufficiently strong mixture of water and ethylene glycol were to be spilled onto a hot exhaust system, then there would be a high risk of fire.

Instrumentation of the Cooling System

When an engine is fitted to a test bed, its systems must be instrumented for the relevant information to be measured. Within the cooling system, the measurements of interest to the test project include some of the following:

- **Coolant temperature**—Into the engine, out from the engine, bypass circuit, heater circuit. The temperature of the coolant is measured using thermocouples.

- **Coolant pressure**—Into the engine, pump, out from the engine, bypass circuit, heater circuit. The pressure within the cooling system is measured using either dial gauges or, if a more accurate measurement is required, the system can be linked by pressure pipes to a transducer.

- **Coolant flow**—Into the engine, pump, out from the engine, bypass circuit, heater circuit. The flow rate of the coolant through the system can be measured using flow turbines fitted to the pipe-work of the system. The positioning of the instrumentation normally is as close as possible to the engine without disrupting the manufacturer's production layout. In some cases, the engine and/or its components may need to be stripped, either partially or wholly, reworked or modified to accept sensors and measuring devices, and rebuilt to ensure that the instrumentation is fitted securely and will not affect the engine or its performance in any way. For frequent testing on a number of engines within a range or batch, it may be preferable to have a complete set of instrument-prepared components that are fitted to the test engine for the test and then are removed after the test for fitting to the next engine.

Health and safety considerations—Personal protection must be worn when dealing with both used and unused coolant. Also, correct disposal of the used coolant must be observed for health and environmental reasons.

Engine Induction System

The purpose of the induction system of the engine is to bring atmospheric air into the engine cylinders to mix it with fuel for the combustion process. During the passage of air to the engine cylinder, the air must be purged of any dirt and foreign bodies that may cause internal damage to the finely machined internal parts of the engine. The point in the induction process at which the air and fuel are mixed varies, depending on the nature of the engine fueling system. In fuel carburetor engines, the air and fuel are mixed at the point where the air inlet tract passes across the fuel jet openings of the carburetor; this actually occurs within the body of the carburetor and upstream from the inlet manifold.

Induction System Components

The following components are part of the engine induction system:

- Air inlet tract
- Air filter element
- Throttle butterfly valve
- Carburetor
- Manifold

Air Inlet Tract

The air inlet tract comprises all ducting and any components that are fitted into the ducting between the extreme point at which air for the engine induction is collected, and the engine inlet manifold, where the air/fuel mixture is accumulated prior to entering the engine cylinders. This tract commonly is constructed from high-pressure molded plastic, which means that it is lightweight and that more complex shapes can be made. This allows for better under-bonnet placement. It is essential that the designer takes into account the volume of air that the engine requires while running at high-speed/high-load conditions. Equally, the test technician and engineer also should consider this important factor when installing engines into test cells. The air filter container is an important part of the inlet tract and either is mounted directly onto the carburetor or is connected to the carburetor by a rigid tube. Engine performance is critically dependent on the air inlet passage configuration; therefore, the air filter mounting must conform to the required standards.

Air Filter Element

Usually, this is designed as a disposable item constructed from convoluted paper. The convolutes give the element sufficient strength to maintain its shape. The element is held between the lower container body and the removable cover. It has two rubber gaskets molded to it, which form an airtight seal within the filter container. Other types include the oil-bath type and the metal gauze outer, sponge inner (K&N) washable type of filter element.

If unfiltered air reaches the engine, the following may result:

- Carburetor jet blockage

- Valve seat damage, causing a loss of cylinder compression

- Engine cylinder or piston scoring, causing a loss of cylinder compression or leading to piston-to-cylinder seizure

- Piston ring wear, causing high oil consumption and blow-by

Throttle Butterfly Valve

Normally an integral part of the carburetor, the throttle butterfly valve is a variable opening disc fitted to a cross-spindle positioned to the lower portion of the carburetor body. This valve is connected to the accelerator pedal of the vehicle.

Where the engine is utilizing fuel injectors, the throttle butterfly valve is fitted within the upper portion of the manifold. On systems that utilize engine management electronics (e.g., ECU), the throttle butterfly valve is positioned to the rear of the airflow-metering device. The purpose of the butterfly valve is to control the volume of air flowing through the inlet tract at given engine speed conditions. The power output of the engine depends directly on the mass of the air and fuel charge that the cylinders of the engine actually take in and combust. Therefore, this relatively simple device can have a significant role in the running of the engine.

Carburetor

The carburetor can be a complex component, and with so many variants available in the workplace, it will not be discussed in any great detail. Suffice it to say that the purpose of the carburetor is to mix fuel and air to the correct ratio for any given engine condition. In essence, it is a tapered tube. As the air passes through the narrowed throat, it is forced to speed up. The increased speed of this air over the main carburetor jet causes additional fuel to be drawn into the airstream and atomized, with the fuel/air ratio remaining constant over the range.

Note that with fuel injection for spark ignition engines, more and more engines are fitted with injectors, which negates the requirement for a carburetor. In the case of direct injection (DI) engines, the injectors are fitted into the cylinder head and inject atomized fuel directly into the individual cylinder combustion chambers.

In the case of indirect injection (IDI) engines, the injection of fuel into the airflow takes place farther upstream from the combustion chamber. Where this occurs upstream will determine the type of injector that is used. Single point injectors (SPI) are used upstream of the throttle butterfly valve, whereas multi-point injectors (MPI) are sited in the individual branches of the inlet manifold. Direct injection has the greater operating pressures.

Manifold

Ten factors must be taken into consideration when designing a manifold, as follows:

1. Flow to each cylinder should be as direct as possible.

2. Charge quantities to each cylinder should be equal.

3. An equal mixture strength of uniformly distributed charge is required for each cylinder.

4. Equal aspiration intervals between manifold branch pipes will prevent charge flow interference between cylinders.

5. Designed for minimal wall wetting and collection points for unburned fuel.

6. An amount of ram pressure charging is needed.

7. Use the smallest tract diameter to maintain adequate air velocity at low engine speeds but without hindering volumetric efficiency at high engine speeds.

8. Use the smallest amount of friction in each branch pipe.

9. Allow enough pre-heating for cold starting and warm-up periods.

10. Provide drainage of the heavier liquid fraction of the fuel.

As will be realized from this list of considerations, manifold design is a precise and complex subject. Therefore, it is sufficient to state that the manifold is a means of connecting the engine to the rest of the induction system. At the same time, it is used as a form of heater element to prevent carburetor icing. With the new generations of plastics now available, induction manifolds can be, and indeed are, made from these plastics. The main benefits are that the manifolds can cost less to manufacture, and the finish of the end product is of a higher quality than those made from aluminum.

Should the manifold-to-cylinder-head sealing gasket have an air leak, it will have detrimental effects on the performance of the engine. Weakening of the air/fuel mixture (i.e., overheating of the piston crown) will cause the engine to misfire on one or more cylinders, leading to loss of power.

The air/fuel charge is the source of energy within an engine when it has undergone the combustion process; however, it also has a secondary role, in that it is used to cool the piston crown area, too. If this were not true, then after a period of running, the piston crown would overheat and melt rapidly. A sign of this will be detonation whereby the air/fuel mixture is ignited prior to the designated ignition point by the spark plug. Another cause of detonation is maladjusted ignition timing; therefore, both the timing and the air/fuel ratio must be correct for the engine to run efficiently.

Instrumentation of the Inlet Manifold

Flow meter turbines can be fitted to measure the rate of airflow through the inlet tract.

Thermocouples can be fitted to the same circuit to measure inducted air temperatures. They also can be fitted to any part of the manifold, internally or externally (surface temperature measurement).

Liquid manometers can be used to measure pressure/depression (negative pressure) levels above and below the throttle butterfly valve.

Pressure transducers can be employed with good results and enable automatic data logging.

Health and safety considerations—Care should be taken when handling the manifold, carburetor, or injectors because modern fuels contain many additives which, when burned, will change their chemical makeup. The full implications of these changes are not yet understood completely, and their effects on humans and the environment also are unknown.

Note that with the continual advances in engine technology and associated electronic control systems, this section deals with only the basic principles of the engine induction system.

Engine Exhaust System

The most important element of an engine exhaust system is the exhaust manifold system.

Exhaust Manifold System

The primary purpose of the exhaust manifold is to provide a route for the gases produced by the combustion process to pass to the exhaust system proper. However, the exhaust manifold also assists in another, and much more fundamental, operation: cylinder scavenging. It is possibly the most important device in the process of scavenging spent gases from the combustion area of the engine.

This scavenging is achieved by utilizing the kinetic energy of the outward-going exhaust gases. These are used to set up a compression wave, which is followed by an expansion wave, the effect of which is to cause the gas pressure to be reduced to a depression within the region of the exhaust port. This operates in conjunction with the valve overlap period, so that the outward-bound exhaust gases from one cylinder can be utilized to assist the induction of another cylinder. Again, this subject is quite complex because each engine configuration requires separate consideration, and much work using airflow rigs is performed to establish exhaust manifold configuration, long before the engine will see a test bed run.

Instrumentation of the Exhaust System

Instrumentation of the exhaust manifold will be similar to that for the induction manifold. An important point here for consideration is that when individual branched exhaust manifolds require instrumentation, great care must be taken to site the sending devices at the same point (e.g., in relation to distance from the exhaust port) for each branch. The test technician also should be aware that significant changes in measured temperatures and pressures will occur as distance from the exhaust port increases. Therefore, it is important to ensure repeatability so that one always positions the thermocouples in the same relative position.

Chapter 12

Quality Standards for Engine Testing

What are the objectives of testing internal combustion engines? Why test them? What does testing provide to the automotive engineer? The answer to all these questions is that testing is a means of comparing differing engine builds, one with another. It is an aid to engine development from Design Level One through production signoff. Testing is the ultimate production quality audit tool. Most importantly, it ensures that the end customer has a product that meets his or her expectations.

If we are going to conduct the test, then it must be done correctly and to the highest standards (i.e., quality testing). There are three considerations when defining quality. It must be ensured that all of the following occur:

- The clients' needs are met.
- The testing is completed at the agreed cost.
- The testing is achieved within the agreed time.

Quality is important to the customer and to the supplier. Without it, the two are unlikely to meet again. In engine testing, quality has huge significance.

Test laboratories are institutions that, of necessity, are built on the principle of total quality. Tests must be conducted to high-quality and pre-determined procedures, using high-quality test control and test vehicle hardware. Otherwise, the test results would be questionable. Because the purpose of engine testing is to produce useful data and information upon which decisions can be made with a high degree of confidence, then it follows that this confidence stems from the careful planning of the tests.

The principal considerations for test technicians and engineers alike are that the engine test must have the following attributes:

- Accuracy
- Consistency
- Repeatability
- Quality workmanship

The customer must receive what he or she is paying for, and clearly everyone involved in the provision of this must understand what the customer is expecting to receive. Indeed, most quality accreditation audits examine the whole operation to establish that all involved are aware of their roles in meeting the outcomes.

Quality Standards for Test Laboratories

Within the test industry, the quality standards that are applicable in other industries have a "stay in business" implication for test laboratories.

For all quality standards to which a company is accredited, the accrediting bodies (e.g., British Standards Institute [BSI], Lloyd's Register Quality Assurance [LQRA], or the United Kingdom Accreditation Service [UKAS]) will ensure compliance by visiting those companies as follows:

- **Surveillance visits every six months**—This will involve the accrediting body visiting to audit parts of the quality system in rotation.

- **Recertification every three years**—This will involve the accrediting body auditing every part of the quality system.

In the United Kingdom, the following quality standards apply. Note that many other internationally accredited quality standards exist, although most follow the same general patterns.

BS EN ISO 9000 Series

The ISO 9000 series sets out methods that can be implemented in an organization to ensure that customers' requirements are met fully. Not only does a fully documented quality management system ensure that the customers' requirements are met, but the organization's requirements also will be met, both internally and externally and at an optimum cost. This is the result of efficient utilization of the available resources—material, human, and technological. This quality standard covers the entire operation of the company, including methods and procedures that a company commits to follow to assure the quality of its service or product that it is supplying to its customers.

The ISO standard is endorsed by Lloyd's Register Quality Assurance (LRQA) or by British Standards Institute (BSI) and is an international standard. The standard covers disciplines such as the following:

- Customer initial contact through to invoicing procedures
- Service or product tracking through the company procedures
- Service or product quality control procedures
- Quality auditing systems
- Customer concern, follow-up procedures, and so forth

When a company is accredited to BS EN ISO 9000, it is for the entire engine test operation.

United Kingdom Accreditation Service (UKAS) (EN 45001)

The United Kingdom Accreditation Service (UKAS) is recognized by the government of the United Kingdom as the body responsible for the assessment and accreditation of organizations performing testing. UKAS was formed in 1995 as a company limited by guarantee by the merger of the National Measurement Accreditation Service (NAMAS) and the National Accreditation Council for Certification Bodies.

Testing organizations that comply with the requirements of the NAMAS Accreditation Standard and Regulations are granted NAMAS accreditation by UKAS and are entitled to use the NAMAS logo.

This body provides accreditation for companies operating in the engine testing industry, primarily testing engines, fuels, oils, additives, and components by measurement and sampling.

Engine testing establishments apply to UKAS to be accredited for individual test procedures that they perform on their site or premises. These tests are already recognized test procedures; the test house is seeking to demonstrate that it can run the tests competently to UKAS standards.

Implications of Quality Standards

What does this mean for engineers and test technicians? Test laboratories are, of necessity, quality environments because all the work performed is required to be traceable to a baseline of a known standard. For example, consider the following:

- Your test procedures are developed generally to a recognized industry standard and where the test results are required to have UKAS accreditation. Thus, a UKAS Accredited Test Procedure must be followed. This test procedure will be the same in any company wishing to offer that service to a customer seeking UKAS accredited results.

- Calibrated equipment must be used to reduce the risk of introducing faulty data and information into the test results through inaccurate measurements or physical activities such as torque-down procedures.

- Documentation and records must be kept meticulously to record the stages and results of tests. These records provide traceability and are the basis of customer reports, and the documentation will vary, depending on the test. In meeting the client's needs, the passage of information up and down the links within the test team is vital, and it begins for the technician with the test instructions. The technician must understand the client's objectives for the test, how the engineer is proposing to achieve them, and the technician's role in the project.

- Total preventive maintenance (TPM) procedures will prevent the introduction of faults into tests and will help to minimize the causes of unnecessary downtime. Downtime refers to periods when the test has been stopped or interrupted for unforeseen reasons (not adjustments or component changes written into the test procedures).

    Downtime can cost companies at least $1,250 per day income, not to mention engineer and project inconvenience. Often, a five-minute task will prevent escalation into major causes of downtime.

    That is the essence of total preventive maintenance.

- The training of technicians toward a high basic standard is required to service the needs of the test industry, which is operating regularly at or near the cutting edge of engine development.

    Note that this training, combined with an enhanced interest in the developments within the industry, will provide the basis of a good technician/engineer.

- Work practices and fitting skills are required to be of the highest standard in a highly competitive industry. "Right first time" and forward thinking should be the absolute baseline for an engine test/build technician's work.

- Witness reports frequently are generated by assessments conducted as part of the UKAS accreditation "rules" to continuously "monitor the competence of those technicians" who will carry out UKAS accredited test procedures.

  Note that the assessments should be looking at key skill areas within the test procedures (i.e., component replacement skills), test control equipment operation and use, and underpinning knowledge of the test aims and outcomes. An awareness of the rating function will always add value to the technician's engine testing knowledge.

Also worth noting at this stage is that it takes years of hard work to build a good reputation; however, it takes only seconds to destroy it. This applies equally to companies, products, and individuals. There are huge pressures on the automotive engineer that underline the need for quality and calibration.

Chapter 13

Base Calculations

This chapter concentrates on the calculations involved when testing engines. A selection of base calculations is described, with appropriate usage and suitability of the calculations. The chapter starts with some brief descriptions of the most common ones.

Torque Backup

Torque backup is best described as

$$\frac{(\text{Maximum torque} - \text{Torque at maximum speed})}{\text{Torque at maximum speed}}$$

Development of advanced common rail systems and associated equipment has introduced additional freedom to vary fuel delivery over the speed range of an engine. For example, the turbocharger matching process must be linked closely with the fuel system matching, even after optimum injection rates, pressures, nozzle sizes, and swirl have been achieved.

Motoring Mean Effective Pressure

The motoring mean effective pressure (MMEP) is determined from the torque required to motor an engine at a pre-determined condition and can be calculated as

$$\text{MMEP (bar)} = \frac{1000 \times \text{Motoring torque (Nm)} \times \text{Number of strokes}}{\text{Capacity (cc)}}$$

Volumetric Efficiency

Volumetric efficiency $\acute{\eta}_v$ is a measure of the effectiveness of the induction and exhaust processes. It is convenient to define volumetric efficiency as

$$\eta'_v = \frac{\text{Volume of ambient density air inhaled per cylinder per cycle}}{\text{Swept volume per cylinder at ambient pressure and temperature}}$$

Assuming that air obeys the gas laws, this can be written as

$$\eta'_v = \frac{\text{Volume of ambient density air inhaled per cylinder per cycle}}{\text{Cylinder volume (cc)}}$$

$$\eta'_v = \frac{V_a}{V_s N}$$

where

V_a = volumetric flow rate of air with ambient density

V_s = engine swept volume

N = revolutions per second for a two-stroke unit, or revolutions per second divided by 2 for a four-stroke engine

Specific Fuel Consumption

The specific fuel consumption (SFC) can be defined as

$$\text{SFC} = \frac{\text{Mass flow rate of fuel}}{\text{Power output}}$$

Correction Factors

Pressure:

$$P = \frac{F}{A}$$

Force per unit area, where the unit is Pascal (Pa) N/m^2:

1 bar = 10^5 Pa

1 standard atmosphere = 1.01325 bar

1 bar = 14.504 psi (pound force per square inch, 1N = 0.2248 pound force)

1 psi = 6894.76 Pa

Psia versus psig:

a = absolute

g = gauge

psia = psig + 1 atmosphere pressure

For example,

    25 psig = 25 + 14.504 = 39.504 psia

    25 psig = 25 × 6894.76 Pa

        = 172,368 Pa (gauge)

        = 172,368 Pa + 101,325 Pa (1 atmosphere)

        = 273,693 Pa (absolute)

Phase

Phase is the nature of a substance. Matter can exist in three phases: solid, liquid, or gas.

Cycle

If a substance undergoes a series of processes and returns to its original state, then it is said to have been taken through a cycle.

Process

A substance can be said to have undergone a process if the state is changed by operation of that process having been carried out on it.

- Constant temperature process
- Isothermal process
- Constant pressure process
- Isobaric process
- Constant volume process
- Isometric process or isochoric process

Heat

Temperature t (Celsius) = T − 273.15 (Kelvin)

Q – heat energy in joules/kg

Specific heat capacity:

    Heat transfer per unit temperature

$$C = \frac{dQ}{dt}$$

    The unit is joules/kg K (joules per kilogram per Kelvin).

Calorific value is the heat liberated by burning the unit mass or volume of a fuel (e.g., gasoline: 43 MJ/kg).

Work done in a polytropic process:

$$\text{Work done} = \int_{V_1}^{V_2} P\,dV = C\int_{V_1}^{V_2} V^{-n}\,dV$$

$$= \frac{C}{-n+1}\left(V_2^{-n+1} - V_1^{-n+1}\right) = \frac{p_1 v_1 - p_2 v_2}{n-1}$$

Enthalpy

A thermodynamic quantity equal to the internal energy of a system plus the product of its volume and pressure.

$$H = U + PV$$

where

U = internal energy

P = pressure

V = volume

Specific Enthalpy

The specific enthalpy of a working mass is the property of that mass used in terms of dynamics. The S.I. unit for specific enthalpy is joules per kilogram.

$$h = U/m = u + Pv$$

Principle of the Thermodynamic Engine

The engine converts chemical energy into usable power and torque.

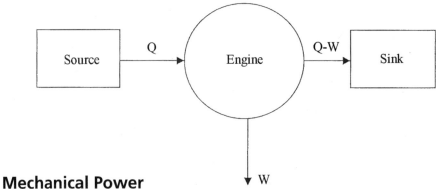

Mechanical Power

That which is transmitted through a mechanical system and used to perform work.

$$\text{Power} = \frac{\text{Work done}}{\text{Time taken}}$$

where the unit is J/s = watt.

Electrical Power

The product of voltage and current.

$$W = IV$$

where the unit is J/s = watt.

Laws of Thermodynamics

The Zeroeth Law: If bodies A and B are in thermal equilibrium, and bodies A and C are in thermal equilibrium, then bodies B and C must be in thermal equilibrium.

The First Law:

$$W = Q$$

This means that if some work W is converted to heat Q, or some heat Q is converted to work W, then W = Q. It does not mean that all work can convert to heat in a particular process.

The Conservation of Energy

For a system,

$$\text{Initial energy} + \text{Energy entering} = \text{Final energy} + \text{Energy leaving}$$

$$\text{Potential energy} = gZ$$

$$\text{Kinetic energy} = \frac{1}{2}mC^2$$

Joule's Law

Internal energy of gas is the function of temperature only and independent of changes in volume and pressure,

$$P_1 V_1^n = P_2 V_2^n$$

$$\frac{P_1}{P_2} = \left[\frac{V_2}{V_1}\right]^n = \left[\frac{V_1}{V_2}\right]^n \quad \text{and} \quad \frac{V_1}{V_2} = \left[\frac{P_1}{P_2}\right]^{\frac{-1}{n}}$$

The specific heat capacity at constant volume is C_v.

$$U_2 - U_1 = mc_v(T_2 - T_1) \quad \text{(change in internal energy)}$$

The specific heat capacity at constant pressure is C_p.

$$U_2 - U_1 + P(V_2 - V_1) = mc_p(T_2 - T_1)$$

From the characteristic equation

$$\frac{P_1 V_1}{T_1} = \frac{P_2 V_2}{T_2}$$

$$\frac{T_1}{T_2} = \frac{P_1 V_1}{P_2 V_2} = \left(\frac{V_2}{V_1}\right)^n \frac{V_1}{V_2} = \left(\frac{V_2}{V_1}\right)^{n-1}$$

and

$$\frac{T_1}{T_2} = \frac{P_1 V_1}{P_2 V_2} = \frac{P_1}{P_2}\left(\frac{P_1}{P_2}\right)^{\frac{-1}{n}} = \left(\frac{P_1}{P_2}\right)^{\frac{n-1}{n}}$$

that is,

$$\frac{T_1}{T_2} = \left(\frac{V_2}{V_1}\right)^{n-1}$$

$$\frac{T_1}{T_2} = \left(\frac{P_1}{P_2}\right)^{\frac{n-1}{n}}$$

Entropy

A measure of the unavailability of a system's energy to do work.

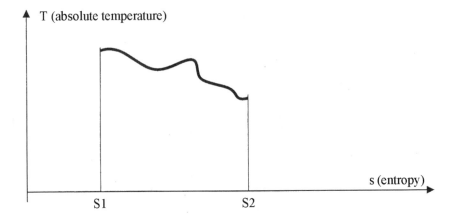

Revisable heat transfer is Q_{rev}

$$dQ_{rev} = T ds$$

$$Q_{rev} = \int_{s_1}^{s_2} T ds$$

An isolated system can change to states only of equal or greater entropy.

Entropy for gas:

$$\text{Polytropic process} = PV^n = C$$

$$\text{Heat transferred} = dQ = \frac{\bar{c}-n}{\bar{c}-1} Pdv \qquad \bar{c} = \frac{c_p}{c_v}$$

$$ds = \frac{dQ}{T} = \frac{\bar{c}-n}{\bar{c}-1}\frac{P}{T}dv = \frac{\bar{c}-n}{\bar{c}-1}\frac{R}{v}dv$$

(combined characteristic, e.g., where s minus the specific entropy v is a unit of V)

$$\int_{s_1}^{s_2} ds = \frac{\bar{c}-n}{\bar{c}-1} R \int_{v_1}^{v_2} \frac{dv}{v} \qquad s_2 - s_1 = \frac{\bar{c}-n}{\bar{c}-1} R \ln \frac{v_2}{v_1}$$

$$s_2 - s_1 = c_v \frac{\bar{c}-n}{n-1} \ln \frac{T_1}{T_2}$$

$$= c_v \frac{\frac{c_p}{c_v} - n}{n-1} \ln \frac{T_1}{T_2} = \frac{c_p - c_v n}{n-1} \ln \frac{T_1}{T_2}$$

$$= \frac{-c_p(n-1) + c_p n - c_v n}{n-1} \ln \frac{T_1}{T_2}$$

$$= -c_p \ln \frac{T_1}{T_2} + (c_p - c_v) \frac{n}{n-1} \ln \frac{T_1}{T_2}$$

$$= c_p \ln \frac{T_2}{T_1} - R \ln \left[\frac{T_1}{T_2}\right]^{\frac{n-1}{n}} \qquad (c_p - c_v = R)$$

that is,

$$s_2 - s_1 = c_p \ln \frac{T_2}{T_1} - R \ln \frac{P_2}{P_1}$$

For example, the air at T = 288 K and standard atmosphere is compressed to P = 5 times of standard atmospheric pressure with a temperature of 456.1 K. Calculate the entropy values.

For air, C_p = 1005 J/kgK and R = 287 J/kgK.

Before compression,

$$s_2 = 1005 \ln \frac{288}{273.15} - 287 \ln \frac{0.101}{0.101} = 53$$

After compression,

$$s_2 = 1005 \ln \frac{456.1}{273.15} - 287 \ln \frac{5 \times 0.101}{0.101} = 53$$

There is no change in entropy.

The pressure, humidity, and temperature of the ambient air inducted into an engine, at a given speed, affect the air mass flow rate and hence the potential power output. Correction factors are used to adjust measured wide open throttle power to standard atmospheric conditions to provide a more accurate basis for comparison among engines.

The basis for these correction factors is the equation for one-dimensional steady compressible flow through an orifice.

Correction Formulae

- 88/195/EEC
- EEC80/1269
- ISO 1585
- JIS D 1001
- SAE J1349
- DIN 70 020

Correction formulae: $88/195/EEC = \left[\dfrac{99}{P}\right]^{1.2} \times \left[\dfrac{T+273}{298}\right]^{0.6}$

Pressure: Standard atmosphere   1.01325 bar

Temperature: 20°C (293°K)

Corresponding density: ρ_s = 1.205 kg/m³

where

      P = barometric pressure (atmospheric) − partial vapor pressure (kPa)

      T = ambient intake air temperature in degrees Celsius (°C)

      ρ_s = density of air

The standard conditions are considered to be a barometric pressure of 99 kPa and a temperature of 298°K (25°C).

The partial vapor pressure required to calculate the dry barometric pressure term P in the equation 88/195/EEC can be obtained in two ways.

1. Using wet and dry air bulb temperature measurements and psychometric tables

2. Using relative humidity and ambient air temperature measurements and the following equations:

   Dry barometric pressure = Atmospheric pressure − Partial vapor pressure

$$P = Pa - Ppv$$

when

$$Ppv = \frac{\text{Relative humidity } (\%) \times \text{Saturated vapor pressure (kPa)}}{100}$$

The saturated vapor pressure Psv (kPa) is given by

$$\text{Log}^{10} Psv = \frac{30.59051 - 8.2 \times \log^{10}(T) + (2.480E - 3) \times T - (3.142 \times 31)}{T}$$

where

$$T(k) = Ta(°C) + 273.15$$

Ta = ambient air temperature

Correction formulae (DIN):

$$\text{Correction factor} = 760(273 + T/P)_{293}$$

where

$$760 \left[\frac{273 + T}{P + 293} \right]$$

P = barometric pressure (mm Hg)

T = ambient intake air temperature (°C)

Examples of Calculations Required Within the Test Cell Environment

Test Bed Fuel Flow Measurement

Many test beds are equipped with gravimetric fuel measuring devices that provide an output in the form of mass of fuel used per selected time interval (e.g., grams of fuel consumed in 30 seconds).

Test Bed Airflow Measurement

To calculate the air/fuel ratio, we need to know the fuel flow and the air used by an engine. This can be done chemically using gaseous emission measuring equipment, but a cross-reference is a prerequisite of intelligent testing.

The actual measurement of air mass flow rate is not carried out on a regular basis because it can be calculated from SPINDT AFR (exhaust gas) and from brake-specific fuel consumption. However, various airflow meters are available for measuring engine air consumption, namely, the following:

- Lucas Dawe hot wire corona discharge
- Hot wire wheat stone bridge balance system
- Alcock viscous airflow meter
- Base orifice plate

All types of measuring devices should be used upstream of any intake pressure pulsations to allow for stable readings to be taken and logged. (This usually means prior to an air cleaner/resonator volume.)

Brake Specific Fuel Consumption

Having measured the fuel mass flow rate by one of the methods highlighted in this book, a more useful parameter—the brake specific fuel consumption (BSFC)—can be calculated. It is defined as

$$BSFC = \frac{\text{Fuel mass flow rate}}{\text{Measured brake power P}} = \frac{mf}{P}$$

with units of grams of fuel consumed to deliver 1 kW over 1 hour (g/kWh).

The BSFC provides a measure of how efficiently an engine is using fuel supplied to produce work. For spark ignition engines, typical values of BSFC at wide open throttle are between 250 and 260 g/kWh.

Figure 13.1 gives an example of a specific fuel consumption contour map for the complete operating range of an engine. The contour lines represent lines of constant specific fuel consumption; the lower the figure, the more efficient the running condition.

Brake specific fuel consumption also can be applied to other items whose output is compared with the brake power delivered by the engine. For example, in emissions, one will note brake specific hydrocarbons (BSHC).

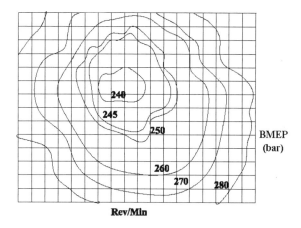

Figure 13.1 Oyster curve example.

Brake Specific Air Consumption

Similarly, brake specific air consumption (BSAC) is defined as the air mass flow rate per unit power output

$$BSAC = Airflow/power \text{ (units kg/kWh)}$$

where power is the brake power output. This provides a measure of how efficiently an engine is using the air supplied to produce work.

Factors that affect the BSFC and BSAC are as follows:

- Compression ratio
- Air/fuel ratio and ignition settings
- Friction—Rubbing in the engine and accessories
- Pumping losses—Intake system restrictiveness/cylinder head design and exhaust system design
- Calorific value of the fuel
- Barrel swirl ratio
- Heat pickup through the induction system
- Heat transfer from the combustion chamber
- Mixture preparation
- Fuel/air mixture distribution

Efficiencies

How much energy reaches the flywheel (or dynamometer) compared to how much theoretically could be released is a function of three efficiencies, namely, the following:

1. Thermal efficiency
2. Mechanical efficiency
3. Volumetric efficiency

Thermal Efficiency

Thermal efficiency can be quoted as either brake or indicated. Indicated efficiency is derived from measurements taken at the flywheel. The thermal efficiency sometimes is called the fuel conversion efficiency because it is defined as the ratio of the work produced per cycle to the amount of fuel energy supplied per cycle that can be released in the combustion process. The available fuel energy is obtained by multiplying the mass of fuel supplied by the heating value of the fuel, as

$$\eta t = \frac{Wc}{m_f Qhv} = \frac{P}{\dot{m}f Qhv}$$

where

Wc = work per cycle

m_f = mass of fuel inducted per cycle

Q_{hv} = heating value of the fuel

P = power output

$\dot{m}_f$ = fuel mass flow rate

Because specific fuel consumption (SFC) can be expressed as

$$SFC = \frac{\dot{m}_f}{P}$$

the equation for ηt can be rewritten as

$$\eta t = \frac{1}{SFC \times Qhv} 1/SFC = \frac{3600}{SFC(g/kWh) \times Qhv(MJ/kg)}$$

because Qhv for gasoline = 43.5 MJ/kg.

Therefore, the brake thermal efficiency is 82.76.

Mechanical Efficiency

The mechanical efficiency compares the amount of energy imparted to the pistons as mechanical work in the expansion stroke to that which actually reaches the flywheel or dynamometer. Thus, it is the ratio of the brake power delivered by an engine to the indicated power

$$\eta m = \frac{\text{Brake power}}{\text{Gross indicated power}} = \left[\frac{\text{bmep}}{\text{imep gross}}\right]$$

Also,

Brake thermal efficiency = Indicated thermal efficiency × Mechanical efficiency

Volumetric Efficiency

The parameter used to measure the ability of an engine to breathe air is the volumetric efficiency, ηv. It is defined as the ratio

$$\eta v = \frac{\text{Mass of air inducted per cylinder per cycle}}{\text{Mass of air to occupy swept volume per cylinder at ambient pressure and temperature}}$$

Air/Fuel Ratio

The air/fuel ratio (AFR) mixture induced into the cylinder will ignite properly and combust only if the air/fuel ratio lies within a certain range. The normal operating range for a naturally aspirated spark ignition engine is between 12 and 18:1 AFR.

Note that combustion limits for gasoline/air mixtures theoretically are 3:1 to 40:1 but practically are nearer to 9:1 to 25:1.

The stoichiometric air/fuel ratio is defined as the mass of air necessary to completely combust a mass of fuel (i.e., there is just enough oxygen for conversion of all the fuel into completely oxidized products).

The stoichiometric AFR is dependent on the composition of the fuel. (Typically for gasoline, it is 14.5:1.) For this reason, two additional parameters are used for defining mixture composition

$$\text{Excess air factor } \lambda = \frac{(A/F) \text{ actual}}{(A/F) \text{ stoichiometric}}$$

$$\text{Fuel-to-air equivalence ratio } \theta = \frac{(F/A) \text{ actual}}{(F/A) \text{ stoichiometric}}$$

Table 13.1 shows some typical air/fuel ratios from rich to lean. (Rich mixtures are less than λ 1.)

TABLE 13.1
AIR/FUEL RATIO VERSUS LAMBDA

Rich			⟶	Lean
AFR	12:1	14:1	16:1	18:1
λ	0.83	0.97	1.10	1.24
θ	1.21	1.04	0.91	0.81

The air/fuel ratio can be calculated by measuring the air and fuel mass flow rates. However, in most instances, the air/fuel ratio is measured by exhaust gas analysis. From the relative concentrations of exhaust pollutants, the air/fuel ratio can be calculated from the SPINDT equation.

First Law of Thermodynamics

When a system undergoes a thermodynamic cycle, then the net heat supplied to the system from its surroundings is equal to the net work done by the system on its surroundings. This is based on the conservation of energy principle and in symbols can be represented as

$$\sum dQ = \sum dW$$

A system operating in a cycle and producing a net quantity of work from a supply of heat is called a heat engine

$$Q_1 - Q_2 = W$$

where

Q_1 = heat supplied

Q_2 = heat rejected

W = net work done

The thermal efficiency of a heat engine is defined as the ratio of net work done in the cycle to the gross heat supplied in the cycle

$$\eta = \frac{W}{Q_1} \qquad (13.1)$$

Substituting $Q_1 - Q_2 = W$ gives

$$\eta = \frac{Q_1 - Q_2}{Q_1} = 1 - \frac{Q_2}{Q_1} \qquad (13.2)$$

Second Law of Thermodynamics

Although the net heat supplied in a cycle is equal to the net work done, the gross heat supplied must be greater than the net work done. That is,

$$Q_1 > W$$

and hence the thermal efficiency of a heat engine will be less than 100%.

Consider the Otto cycle. By calculating the heat supplied and rejected in the cycle, its thermal efficiency can be found. The heat flow in a reversible constant volume process is given by

$$Q = mCv(\Delta T)$$

where

    m = mass

    ΔT = temperature difference

    Cv = specific heat at a constant volume (heat required to raise the unit mass through one degree temperature rise)

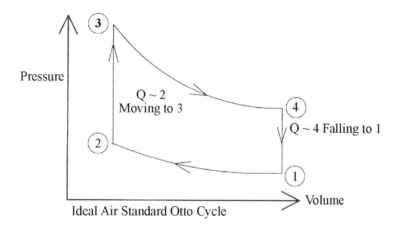

Figure 13.2 Basic pressure volume curve.

Thus, referring to the Otto cycle PV diagram again, the heat supplied Q2 through to 3 at constant volume between temperatures T_2 and T_3 (per kilogram of air) is given by

$$Q(2 \text{ to } 3) = Cv(T_3 - T_2) \qquad (13.3)$$

Similarly, the heat rejected Q4 through to 1 at constant volume between temperatures T_4 and T_1 (per kilogram of air) is given by

$$Q(4 \text{ to } 1) = Cv(T_4 - T_1) \qquad (13.4)$$

Because the compression and expansion phases are adiabatic (i.e., involve no heat transfer), Eqs. 13.3 and 13.4 contain the required terms to substitute into Eq. 13.2 as

$$\eta = 1 - \frac{Cv(T_4 - T_1)}{Cv(T_3 - T_2)}$$

$$\eta = 1 - \frac{(T_4 - T_1)}{(T_3 - T_2)}$$

Also, because the compression and expansion phases are isentropic processes (i.e., reversible and adiabatic), the following relationships apply

$$\frac{T_2}{T_1} = \left[\frac{V_1}{V_2}\right]^{\gamma-1} \quad \text{and} \quad \frac{T_3}{T_4} = \left[\frac{V_4}{V_3}\right]^{\gamma-1}$$

where $\gamma = \frac{C_p}{C_v}$.

Because $V_4 = V_1$ and $V_2 = V_3$,

$$\frac{T_2}{T_1} = \frac{T_3}{T_4} = \left[\frac{V_1}{V_2}\right]^{\gamma-1}$$

Now,

$$\frac{V_1}{V_2} = \frac{\text{Swept volume} + \text{Clearance volume}}{\text{Clearance volume}} = \text{Compression ratio}$$

Thus, it can be seen that the thermal efficiency of the Otto cycle depends on only the compression ratio.

Mathematical Basis for Power Correction Factor

The power output for a given engine type is dependent largely on the air mass flow rate that it achieves. This will vary as the pressure, humidity, and temperature of the air it inducts changes.

The basis for the correction factors used to standardize power and volumetric efficiencies is the equation for one-dimensional steady compressible flow through an orifice or a flow restriction of effective area (A_E):

$$\overset{\circ}{m} = \frac{A_E P_O}{\sqrt{R T_O}} \left\langle \frac{2\gamma}{\gamma - 1} \left[\left(\frac{P}{P_O}\right)^{\frac{2}{\gamma}} - \left(\frac{P}{P_O}\right)^{\frac{(\gamma+1)}{\gamma}} \right] \right\rangle^{\frac{1}{2}}$$

The derivation of this equation assumes that the fluid is an ideal gas, with gas constant R, and the ratio of specific heats $\gamma = \frac{C_O}{C_V}$ is a constant, P_O and T_O are the total pressure and temperature upstream of the restriction, respectively, and P is the pressure at the throat of the restriction.

If, in the engine, $\frac{P}{P_O}$ is assumed constant at wide open throttle, then for a given intake system and engine, the mass flow rate of dry air m² varies as

$$m^2 \alpha \frac{P_O}{\sqrt{T_O}}$$

Thus, to compare power outputs from days with changing ambient conditions, the actual pressures and temperatures must be related to a set of standard conditions. Therefore, if we write $ma_{std} = CF \times ma\,(measured)$, then using the preceding relationship between mass airflow rate and pressure/temperature, the correction factor (CF) can be expressed as

$$CF = \frac{ma_{standard}}{ma_{measured}} = \frac{\left[\frac{P}{\sqrt{T}}\right]_s}{\left[\frac{P}{\sqrt{T}}\right]_m} = \frac{P_{standard}}{P_{measured}} \left[\frac{T_{measured}}{T_{standard}}\right]^{1/2}$$

Because $P_{measured}$ here is assumed to be the dry air pressure, the pressure of any water vapor needs to be subtracted from the actual measured air pressure. Then the preceding expression can be written as

$$CF = \frac{Psd}{Pm - Pvm} \left[\frac{Tm}{Ts}\right]^{1/2}$$

where

    Psd = standard dry air absolute pressure

    Pm = measured ambient air absolute pressure

    Pvm = measured ambient water vapor partial pressure

    Tm = measured ambient temperature

    Ts = standard ambient temperature

The basic principle of the internal combustion engine is that the rapid burning of a combustible mixture produces a release of energy in the form of a pressure rise within the cylinder. This energy is harnessed by various means and is converted into rotary motion. The formulae presented here are those commonly used in the management of air and fuel and the measurement of power and torque. Before going any further, I will review some terms commonly used when discussing engine performance.

Swept Volume

At top dead center, the volume remaining above the piston is termed the clearance volume. The swept volume is defined as the volume above the piston at bottom dead center less the clearance volume.

Hence,

$$\text{Swept volume} = \text{Total volume} - \text{Clearance volume}$$

Compression Ratio

The compression ratio (CR) is the ratio of the volume between the piston and the cylinder head before and after compression.

$$CR = \frac{\text{Total} + \text{Clearance volume}}{\text{Clearance volume}} = \left[\frac{\text{Swept volume}}{\text{Clearance volume}}\right] + 1$$

Typical values for the compression ratio are 8:1 to 12:1 for spark ignition engines and up to 24:1 for compression ignition engines.

Brake Mean Effective Pressure

The calculated brake mean effective pressure (BMEP) is measured in bar, kilograms per square meter (kg/m^2), or pounds per square inch (psi) as

$$BMEP = \frac{\text{Brake work output (Nm) per cylinder per cycle}}{\text{Swept volume per cylinder } (m)^3}$$

The brake mean effective pressure is a better measure of engine performance than power, kilowatts (kW), or brake horsepower (BHP) for comparative purposes because it does not depend on engine size (Table 13.2).

**TABLE 13.2
TYPICAL BMEP**

Engine Type	Peak Torque BMEP Bar	Peak Power BMEP Bar
Spark ignition N/A	8.2 ~ 12.0	6.5 ~ 11.0
Spark ignition T/C	12.5 ~ 19.0	9.0 ~ 16.0

Pumping Mean Effective Pressure

The pumping mean effective pressure (PMEP) is a measure of the work done in drawing a fresh mixture through the induction system into the cylinder and expelling the burned gases out of the cylinder and through the exhaust system.

Compression/Expansion Mean Effective Pressure

In effect, the compression/expansion mean effective pressure (CEMEP) is the same as the gross IMEP.

Friction Mean Effective Pressure

The friction mean effective pressure (FMEP) is a measure of rubbing friction work and accessory work:

$$\text{IMEP gross} = \text{IMEP net} + \text{PMEP}$$

$$\text{CEMEP} = \text{IMEP net} + \text{PMEP}$$

$$\text{BMEP} = \text{IMEP gross} - \text{PMEP} - \text{FMEP}$$

Brake Torque and Power

Power is the product of torque and angular velocity. Torque is a measure of the ability of an engine to do work, and power is the rate at which work is done

$$P(\text{power}) = T(\text{torque}) \times w(\text{engine speed})$$

where

w = angular velocity in radians per second

2π radians = 360° of crankshaft rotation

$$P = \frac{2\pi NT}{60} = \frac{T \times N}{9550}$$

when N is the engine speed in revolutions per minute.

1 kW = 1.3596

PS = 1.341 BHP

1 Nm (torque) = 0.7376 lbf/ft = 9.81 kgf./m and 1 lbf/ft = 1.3558 Nm

$$\text{BMEP} = \frac{1200 \times P}{V \times N}$$

when

T = torque (Nm)

V = capacity (liters)

P = power (kW)

N = engine speed, rev/min

Also, the units of BMEP are in bar (1 bar = 100 KPa).

Motoring Mean Effective Pressure

The motoring mean effective pressure (MMEP) is determined from the torque required to motor an engine at a predetermined condition and can be calculated as

$$\text{MMEP (bar)} = \frac{1000 \times \text{Motoring torque (Nm)} \times \text{Number of strokes}}{\text{Capacity (cm)}^3}$$

Glossary of Terms and Acronyms Used in the Design and Testing of Internal Combustion Engines

AAFI—Air-assisted fuel injection.

ABI—Advanced break-in.

Absorbed horsepower—Total horsepower absorbed by the absorption unit of the dynamometer and the friction components of the dynamometer.

Accelerated life test—Any set of test conditions designed to reproduce in a short time the effects encountered in the field.

Actual horsepower—The actual horsepower is the load horsepower that includes the friction in the dynamometer bearings and inertia simulation mechanism.

"A" curve—Maximum performance test.

Additive removal test—A test conducted by recirculating a specific additive type of oil through a test element for a specific period of time.

Aeration—The entrapment of gas in the coolant or oil.

After-boil—Boiling of the coolant after engine shutdown, caused by residual heat.

Air cleaner element—All airflow measurements are corrected to 25°C at 100 kPa standard test condition.

Air density—1.2250 kg/m^3.

Air viscosity—1.7894×10^5 Ns/m^2.

Ambient—Used to denote surroundings, such as ambient temperature.

AQL—Acceptable quality level.

ATB—Air-to-boil temperature; the ambient temperature at which the coolant at the radiator reaches its boiling point.

Auto—Automatic.

Available secondary spark voltage—The minimum voltage at the spark terminal with the terminal open circuited and insulated from ground.

BAP—Barometric absolute pressure.

Basic engine—A runnable engine equipped with built-in accessories, fuel pump, oil pump, coolant pumps, emission control equipment, and all standard equipment as clearly defined by the individual manufacturer.

Bathtub curve—A bathtub-shaped plot of failure rate of an item (whether reparable or not) versus time; the failure rate initially decreases, then remains reasonably stable, and then begins to rise rapidly.

"B" curve—Production engine performance test.

Beachmarks—A term used to describe the characteristic fracture markings produced by fatigue crack propagation.

Beer-Lambert Law—For the purpose of diesel smoke measurement, an equation expressing the relationship between the opacity of a smoke plume, the effective optical path through the plume, and the opacity of the smoke per unit path length when:

$$K = \text{attenuation coefficient (or actination)}$$
$$L = \text{path length through smoke in meters}$$
$$\text{Opacity} = 1 - e^{-kl}$$
$$N = 100(1 - e^{-kl})$$
$$K = -1/L \, la(1 - N/100) \text{ when L is expressed in meters}$$

Best power—Power at maximum torque achievable by a given test multi-position small engine at the maximum continuous corrected net brake power speed.

Best straight line—A line midway between two parallel straight lines closest together and enclosing all output values on a calibration curve.

BI—Break-in.

BIPO—Break-in and pass-off procedure (run-in).

BLA—Bottom limit advance.

Black smoke—Particles composed of carbon (soot), usually less than 1 µm in size.

BLD—Borderline detonation.

BMEP—Brake mean effective pressure.

BMEPc—Brake mean effective pressure corrected to a constant factor.

Boost pressure—The pressure of the charge of air as it leaves the turbocharger, supercharger, or other compressor.

Breakdown voltage (sparking plug)—The voltage at which a disruptive discharge takes place through or over the surface of the insulation.

Btu—British thermal unit.

°C—Degrees Centigrade.

Calibration curve (emissions)—A line drawn through at least seven points established by calibration gases, which determines the sensitivity of the analytical instrument to unknown gas concentrations.

Calibration traceability—The relation of a transducer calibration, through a specified step-by-step process, to an instrument or group of instruments calibrated by a national standards authority for each country.

Catalyst—A substance that accelerates a chemical reaction but which itself undergoes no permanent chemical change. For automotive emission purposes, catalysts are classified as oxidation catalysts (oxidize HC and CO) and reduction catalysts (reduce NOx) or three-way catalysts (oxidize HC and CO).

"C" curve—Performance as installed on the vehicle.

CFM—Cubic feet per minute.

Charging efficiency—The mass of delivered air retained in the cylinder (displaced volume times the ambient density).

Chemiluminescent (CL) analyzer—An instrument that measures nitric oxide by measuring the intensity of chemiluminescent radiation from the reaction of nitric oxide with ozone. The addition of a converter permits the measurement of oxides of nitrogen.

Cleanable element—A filter element, which when clogged can be restored by a suitable cleaning process to an acceptable percentage of its original flow/pressure differential characteristics.

Closed loop—A control system implementation in which information on output parameters is used to improve the system accuracy or response.

Closed-loop circuit—Feedback system for controlling the air/fuel ratio (AFR).

Closed nozzle—A nozzle incorporating either a poppet valve or a needle valve loaded to open at some predetermined pressure; also a special type of needle valve wherein an integral pintle valve projection from the lower end of the needle is so formed as to influence the rate and shape of the fuel spray pattern during operation.

Closed nozzle valve—A needle valve having two diameters, the smaller at the valve seat. The fuel injection pressure acting on a portion of the total valve area lifts the valve at a pre-determined pressure and then acts on the total area, while the end opposite of the valve seat is never subjected to injection pressure.

CO—Carbon monoxide.

CO_2—Carbon dioxide.

Coast down—The procedure used to determine the total horsepower absorbed by a chassis dynamometer at 50 mph; the time required for the chassis rollers to coast down from 50 mph to a pre-determined speed is noted, and from this, the rolling inertia forces can be calculated.

Coefficient of rolling resistance—The ratio of the rolling resistance to the vertical load.

Coefficient of variation—The standard deviation divided by the mean, multiplied by 100 and expressed as a percentage.

Coil interruption current—The peak current flowing through the coil primary winding at the time of interruption.

Cold start injector (CSI)—An auxiliary fuel injector that supplies additional fuel during cold cranking. Fuel injection generally is continuous, and the fuel rate is based on the orifice size, fuel pressure, and opening duration.

Combined life and partial—A filter life test in which specified amounts of retention test inorganic classified contaminants are admitted to the influent stream at specified intervals during the test to determine the particle separation efficiency of the filter at various stages of clogging.

Combustion chamber—An area of chamber divided by volume with the piston at top dead center.

Compression ratio—The maximum cylinder volume divided by the minimum cylinder volume.

Compressor efficiency—Isentropic total enthalpy rise across the compressor stage divided by the actual total enthalpy rise across the compressor stage.

Compressor pressure ratio—Outlet air absolute pressure divided by inlet air total absolute pressure.

Confidence coefficient—A measure of assurance that a statement based on statistical data is correct; the probability that an unknown parameter lies within a stated interval or is greater or less than some stated value.

Confidence level—Equals $1 - \alpha$, where α is the percent of risk.

Constant failure rate—A term characterizing the instantaneous failure rate in the middle of a bathtub curve or the useful life of a bathtub curve model of an item life.

Constant velocity joint—A universal joint that transmits rotational motion and angular velocity ratio of unity between the output and input shafts.

Constant volume sampler (CVD)—A device for collecting samples of diluted exhaust gas, where the exhaust gas is diluted with air in a manner that keeps the total flow rate of exhaust gas dilution air constant throughout the test. The device permits measuring mass emissions on a continuous basis and, through the use of a second pump, allows a proportional mass sample to be collected.

Contact pressure—The average pressure exerted by a seal on a shaft. This pressure is calculated by dividing the total lip force by the total lip contact area and sometimes is referred to as the radial pressure.

Continuous injection system—A fuel injection system in which fuel flows continuously from the injectors independent of cylinder events; a variable square law orifice frequently controls the flow rate.

Cool charge air temperature—The temperature of the cooled air entering the engine, commonly referred to as the intake manifold temperature or AIT.

Coolant—A liquid used for heat transfer and normally composed of 50% glycol and 50% water by volume.

CPS—Camshaft position sensor.

Curb weight—The weight of a vehicle in a drive-away condition, filled with at least 90% capacity by weight with fuel, lubricants, coolants, and all standard equipment, but without luggage or passengers.

Cyaniding—A case hardening process in which a ferrous material is heated above the lower transformation range in a molten state containing cyanide to cause simultaneous absorption of carbon and nitrogen at the surface and by diffusion creates a concentration gradient. Quench hardening completes the process.

De-aerating tank—A specially designed tank capable of removing entrained air or combustion gas from circulating coolant.

Decreasing failure rate—A term characterizing the instantaneous failure in the first period of the bathtub curve mode of product life.

Delivered air/fuel ratio—The mass of delivered air divided by the mass of delivered fuel.

Density recovery ratio—The ratio of the charge air density at the engine intake manifold to the air density at conditions of ambient temperature and boost pressure.

Diesel engine—Any compression ignition internal combustion engine using the basic diesel cycle, that is, combustion resulting from the spraying of fuel into air heated by compression.

Diesel smoke—Particles, including aerosols suspended in the gaseous exhaust stream of the engine, that obscure, reflect, or refract visible light.

Dilution factor—Based on a stoichiometric equation for fuel with composition $CH_{1.85}$, the dilution factor as defined as

$$\frac{13.4}{CO_2 + (HC + CO) \times 10^{-4}}$$

Dilution tunnel—One of several types of ducts used to dynamically mix engine exhaust with dilution air in a stationary setup. Conditioned air supplied by a blower is mixed with engine exhaust at a mixing orifice in the upstream end of the tube, while sampling takes place at a downstream location.

Direct fuel injection (DFI)—Delivery of fuel directly into the combustion chamber.

Displacement—The volume required to extend the piston rod to its working stroke; the product of multiplying the area times the length of stroke, usually measured in cubic centimeters or cubic inches.

DOT—A flammable solid that is liable to cause fire through friction or can be readily ignited.

Double cardan universal joint—A near-constant velocity universal joint that consists of two cardan universal joints whose crosses are connected by a coupling yoke with internal supporting and centering means and which has intersecting shaft axes. At the design joint angle and at zero, the instantaneous angular velocity is zero (unity), while at other joint angles, it is near unity.

Drawdown—The quantity of coolant that can be lost before impairing cooling system performance, or grade cooling level, under normal operating conditions.

Driveline—An assembly of one or more driveshafts with provisions for axial movement, which transmits torque and/or rotary motion at a fixed or varying angular relationship from one shaft to another.

Driveline ratio—The crankshaft revolutions per minute divided by the revolutions per minute of traction wheels.

Durab—Durability.

Durability—The probability that an item will operate as specified under stated conditions without a wear-out failure; a special case of reliability.

DVT—Design validation test. A test process to evaluate a design change.

Dynamometer—An energy-absorbing device designed to allow measured dynamic operation of the drivetrain of a vehicle while the vehicle remains stationary (chassis dynamometer).

EBP—Exhaust backpressure. The backpressure in an exhaust system.

Eddy currents—Those currents that exist as a result of voltages induced in a body of a conducting mass by a variation of magnetic flux. Note that the variation of magnetic flux is the result of a varying magnetic field or of a relative motion of the mass with respect to the magnetic field.

Electromagnetic compatibility (EMC)—The capability of electronic equipment of systems to be operated in the untended operational electromagnetic environment at designed levels of efficiency.

Electronic control unit (ECU)—An electronic module, one function of which is to calculate a command signal for the injector drive circuit, based on inputs from engine operating sensors.

Electronic fuel injection (EFI)—A general term referring to any fuel injection system in which fuel metering is controlled electronically. The quantity of fuel delivered is scheduled by an electronic control unit, and its output signal is based on information received from several sensors that monitor the operating conditions of the engine.

Excess air factor—The excess air factor can be expressed as

$$\frac{\text{Trapped air - fuel ratio}}{\text{Stoichiometric ratio – air/fuel ratio}}$$

Exhaust gas oxygen sensor (EGOS)—A sensor located in the exhaust system that provides an electrical output that indicates oxygen content.

Exhaust gas recirculation (EGR)—A system that returns a portion of the exhaust gases to the combustion chamber, with the resulting lower combustion temperatures in turn reducing the formation of oxides and nitrogen.

°F—Degrees Fahrenheit.

Flame ionization detector (FID)—An analytical instrument used for determining the carbon concentration of hydrocarbons in a gas sample; a hydrogen-air diffusion flame detector that produces a signal proportional to the mass flow rate of hydrocarbons entering the flame per unit of time.

FMEP—Friction mean effective pressure.

Friction power—The power required to drive the engine alone as equipped for the power test. Friction power may be established by a number of methods including hot motoring friction where a friction torque is recorded at wide open throttle at each test speed point on the power test. All readings are taken at the same coolant and oil temperatures; Willans line; motoring tests. (*See* Willans line.)

GPD—General production design.

Hg—Mercury.

H_2O—Water.

Hydrocarbons—All organic materials, including unburned fuel and combustion by-products present in the exhaust, which are detected by a flame ionization detector.

Idle speed control (ISC)—A general term indicating any device or system that provides programmed control of engine idle speed.

IMEP—Indicated mean effective pressure.

Incremental filter efficiency—A method of calculating filter efficiency based on the total amount of contaminant presented to the filter during any specified segment of the test.

Indicated horsepower—The load horsepower that is set on the dynamometer and does not include the friction in the dynamometer bearings or inertia simulation mechanism.

In-line pump—An injection pump with two or more pumping elements arranged in line, with each pumping element serving only one cylinder.

Intake air temperature sensor (IATS)—A sensor that provides an electrical output proportional to the intake air temperature, and typically is mounted within or ahead of any airflow measuring device.

Intake manifold absolute pressure sensor (IMAPS)—A sensor that provides an electrical output proportional to the absolute pressure within the intake manifold downstream of the throttle plate.

K—Constant.

LBT—Leanest fuel for best torque.

Light-off temperature—The temperature at which the conversion energy reaches a given value.

Loggery—Engine test log storage area.

MBT—Minimum spark (ignition advance) for best torque.

MBT-L—Minimum spark for best torque (MBT) retarded to clear borderline detonation.

Mean auto—Average of top and bottom limit advance settings.

Multi-pass oil cooler—An oil cooler that is circuited so that fluid passes either across or through the core more than once.

Net brake power—The measured power of a fully equipped engine.

NMEP—Net mean effective pressure.

Nobel metal catalyst—A catalyst in which the active material is made from a precious metal such as platinum, palladium, rhodium, or ruthenium.

Non-dispersive infrared analyzer (NDIR)—An instrument that, by absorption of infrared energy, selectively measures specific components.

Obs—Observed.

OCT—Oil consumption test.

Original equipment manufacturer (OEM)—A component of the vehicle that was built according to the specification of the vehicle manufacturer and supplied in the vehicle at the time of original purchase of that vehicle can be said to be an OEM-compliant part.

Oxides of nitrogen (NOx)—The sum total of the nitric oxide and nitrogen oxide in a sample, expressed as nitrogen dioxide.

Parts per million (ppm)—Describing fractional defective, parts per million are obtained by multiplying the percent defective by 10,000 (e.g., 0.01% = 100 ppm).

PDG—Product development group.

PMEP—Pumping mean effective pressure.

Port opening—A term referring to a diesel fuel injection pump of the port and helix sleeve metering type in which the timing is determined by the point of the opening of the port by the metering member, corresponding to the nominal end of the pump delivery.

PTO—Power take-off.

PTO log—Record of PTO unit running hours. (PTO refers to the power take-off unit, fitted between the engine and the dynamometer.)

Rare earth catalyst—A catalyst in which the active material is a rare earth element such as lanthanum or cerium. Note that the rare earth elements range in atomic number from 57 to 71.

Rated net power—Engine net power at rated speed, as declared by the manufacturer.

Relative charge—The relative charge can be expressed as

$$\frac{\text{Mass of trapped cylinder charge}}{\text{Displaced volume} \times \text{Ambient density}}$$

Relative humidity—The ratio, expressed as a percentage, of the amount of water present in a given volume of air at a given temperature to the amount required to saturate the air at that temperature.

Scavenging efficiency—The scavenging efficiency can be expressed as

$$\frac{\text{Mass of delivered air retained}}{\text{Mass of trapped cylinder charge}}$$

SFC—Specific fuel consumption.

SG—Specific gravity.

Specific fuel consumption (SFC)—Mass of fuel consumed per unit of work.

TLA—Top limit advance.

Torsional vibration damper—A torsionally tuned mechanical device that generally consists of an inertia ring attached to a drivetrain component by means of an elastomeric inner ring. It is tuned to a specific disturbing frequency.

Total oxides of nitrogen—The sum total of the measured parts per million (ppm) of nitric oxide (NO) plus the measured parts per million (ppm) of nitrogen dioxide (NO_2), expressed as an equivalent mass of NO_2.

Ultraviolet—Radiant energy having wavelengths of 0.4 to 0.04 μm.

Valve overlap—Valve timing events are the valve opening and closing points in the operating cycle, whereas valve overlap describes that part of the cycle in which both the intake and exhaust valves are open.

Willan's line—A method for estimating the friction power of a diesel engine.

Index

Absolute pressure, 45
AC transient dynamometers, 14–15, 18f, 19
Accuracy, 58
Additives, 120, 121
After-treatment considerations, 158, 159t
Air
    combustion or induction, 29–32
    components of, 122
    exchange systems for, 2–3
    humidity of, 31–32
    quality of, 32
    temperature of, 30
Air blast fuel injection, 62
Air cooling, 29, 32–33
Air filter element, 245–246
Air/fuel ratio (AFR), 127–128, 130–131, 265–266
Air inlet tract, 245
Air pressure, 30–31
Air services for test cell, 29–33
Airflow
    sensors for, 53
    test bed measurement of, 262
Aldehydes, 115
Algorithms for heat release, 182–183
Analyzers
    calibration of, 151
    chemiluminescence NO/NOx, 148, 149f
    combustion fast response, 158
    continuous particulate, 143–144
    maintenance of, 151
    non-dispersive infrared (NDIR), 145–146
    NOx, 148
    oxygen, 148–149
    paramagnetic O_2, 148–149
    venting of, 155
Atoms, 122–123
Augsburg and Krupp, 62
Average exhaust pressure, 210
AVL spark plug, 175f, 176

Back flushing, 145
Balanced chemical equations, 127–129
Base calculations, 253–272
Bathtub curve, 79, 80f
Bed/pallet build records, 97
Bellows, 48, 49

Bench tests, combining with field tests, 84–85
Best practices, test cell, 233–235
Bolometers, 45, 46f
Bond energies, 123–124, 127f
Boost pressure, 225, 227, 229
Bosch, Robert, 61, 63, 68
Bottom dead center (BDC), 162
Bourdon tube pressure gauge, 48, 49
Bowman, Peter, 63
Brake mean effective pressure (BMEP), 131–132, 138, 177, 270
Brake power, 16, 178–179
Brake specific air consumption (BSAC), 263
Brake specific fuel consumption (BSFC), 131, 262, 263f, 263
Brake thermal efficiency, 265
Brake torque, 271
Brake work, 103
Breathing, 29
British Standards Institute (BSI), 250
BS EN ISO 9000 Series, 250
BS/BS EN standards, 1
Burn rate, 187
Bypass system, 242

Cable, high-impedance, 210–211
Calibration, 151, 251
    cold start, 202–207
    definition of terms used in, 57–58
    of instrumentation, 56–59
    temperature, 58
    trade-offs in, 200–202
Calibration range, 58
California Air Resources Board (CARB), 117
Cancer-causing agents, 112
Canister oil filter, 240
Capsules, pressure, 48
Carbon, 113–114
Carbon dioxide (CO_2), 111, 117
Carbon monoxide (CO), 111, 115
Carburetor, 246
Cardan shafts, 19
Catalyst
    diesel oxidation, 117–119
    operation of, 137
Catalyst ageing, 78
Catalytic coatings, 120

Ceramic insulator, 107
Cerium oxide, 121
Cetane number (CN), 72
Charge air cooling, 222, 223f, 224f
Charge amplifier, 211
Checking and inspecting, 233
Chemiluminescence NO/NOx analyzer, 148, 149f
Cleaning, 234
Climate change, 113
Closed-loop control system, 38
Coal-dust-burning engine, 62
Coefficient of variation (COV) of IMEP, 188
Cold fouling, 106
Cold junction compensation, 44
Cold junction, 43
Cold plugs, 102f, 105–106
Cold start, 202–207
Combustion, 113–129
    abnormal, 194
    basic, 161–162
    diesel versus gasoline, 130–131
    formation of pollutants during, 129
    incomplete, 194
    non-ideal, 129
    principles of, 68–72
Combustion analysis, 218–219
    features of, 175–180
Combustion diagnostics, types of, 180–186
Combustion efficiency, 217–218
Combustion fast response analyzers, 158
Combustion measurement, 207–208
Combustion or induction air, 29–32
Combustion pressure, 65–66, 176
Combustion variability, 186–216
    causes of, 186
    impact of, 186–187
    quantifying, 188–194
    steps to improve, 190–193
Common rail systems, 62–63
Compression/expansion mean effective pressure (CEMEP), 271
Compression ignition, 61
Compression ratio (CR), 66, 67, 270
    effective, 218–219
    maximizing, 216–217
Compression stroke, 162, 163f
Concentration of particulate, 140
Confidence level, 82–83
Conservation of energy, 257
Constant fill hydraulic dynamometers, 11
Constant pressure turbocharging, 221
Continuous particulate analyzers, 143–144
Continuously regenerating trap (CRT), 75, 120–122
Control charts, 213–215
Control of Substances Hazardous to Health (COSHH), 1
Control systems, 37–40

Coolant, 24, 244
Cooling, air, 29, 32–33
Cooling circuit, raw water, 22
Cooling fan, 243
Cooling plant, 2–3
Copper core center electrodes, 106
Correction factors, 254–255, 268–272
Correction formulae, 260–261
Coupling, suitability considerations for, 19–20
Covalent bonding, 122–123
Crash cooling, 22
Cycle, 255
Cyclic temperature variation, 79
Cylinder head, 242
Cylinders, pressure measurement in, 166–168, 174, 176
    hardware for, 208–209

Daily checks, recommended, 216
Dalton's law of partial pressures, 128–129
Data integrity, 212–213
DC transient dynamometers, 14–15, 18f, 19
Dead time, 39
Dead-weight tester, 58, 59f
Derivative response, 39–40
Diaphragm pressure transducers, 49–50
Diesel, Rudolf, 61–62
Diesel engines
    advantages of, 66–67
    continuously regenerating trap filters for, 75
    direct versus indirect injection in, 131–137
    disadvantages of, 65–66
    fuel consumption in, 67–68
    and international regulations, 73
    introduction to, 61–64
    particulate filters for, 73
    turbocharging of, 221–231
    typical test procedure for, 86–93
    unit injectors for, 65
Diesel fuels, 67, 72
Diesel Motor Company of America, 62
Diesel oxidation catalyst, 117–119
Diesel particulate filters (DPFs), 73–75
Diesel particulate matter (DPM), 115, 118
Diesel traps, 117
Differential imbalance percentage (DIP), 189–190
Dilute sampling, 138–139
Dilution ratio, 140–141
Direct injection (DI), 66, 70–71, 72, 247
    versus indirect injection, 131–137
Distributor rotary pump, 63, 64f
Disturbances, 38–39
Docking rigs, 33, 35f
Documentation, 251
Drivetrain, efficiency loss mechanisms in, 168–170
Dry barometric pressure, 261

Dry basis calculation, 129
Durability, 79, 80
Durability tests, 77, 78–79, 83, 86
Dynamometers, 4–20
    AC or DC transient, 8, 14–15, 18f, 19, 36
    characteristics of, 17–19
    eddy current, 8, 12–13, 17f, 19, 36
    function of, 7
    hydraulic, 7–12, 17f, 18, 36
        constant fill, 11
        variable fill, 11–12
    mechanism of, 15–17
    operating quadrants of, 16, 17t
    operation of, 7–11, 15–17, 36
    types of, 7–8

Earthed junction, 44
Eddy current dynamometers, 12–13, 17f, 19
Effective compression ratio, 218–219
Electrical equipment, placement of, 31–32
Electrical power, 257
Electrons, 122–123
Emissions, 111–113
    constituents of, 115–116, 156
    controlling, 117
    diesel, 114–122
    European regulations for, 121–122
    factors in monitoring, 155–156
    new measuring methods for, 158
    reduction devices for internal combustion engines, 137–138
    regulation of, 116–117
    testing of, 156–157
EN 45001, 250–251
Encoders, 211–212
Energy, conservation of, 257
Energy balance, 3–4, 5f
Engine block, 241–242
Engine life, 79, 85
Engineering records, 97
Engines
    coal-dust-burning, 62
    heat, 266
    optical, 175f, 176
    records on running, 98
    testing of, 77–99
        pointers for, 36–37
        quality standards for, 249–252
    thermodynamic, principle of, 256
    See also Diesel engines, Internal combustion engines
Enthalpy, 256
Entropy, 258–260
Environmental Protection Agency (EPA), 116, 117
Equivalence ratio, 218
Error term, 39
European emissions regulations, 121–122

Excess air factor, 265
Exhaust gas emissions, see Emissions
Exhaust gas recirculation (EGR), 132–133, 138, 158–159, 160t, 192–193
Exhaust stroke, 162t, 165
Exhaust system tests, 78
Expansion tank, 243
Exposed junction, 44

Fault diagnosis, 98–99
FEV ion gap plug, 175f, 176
Feyens, Francois, 63
Filters
    41-point, 198, 199f
    continuously regenerating trap, 75
    diesel particulate (DPFs), 73–75
    handling of, 142
    weighing of, 142
First law of thermodynamics, 257, 266
Fitting practices, 233–235, 251
Flame ionization detector (FID), 147
Flow measurements, instrumentation for, 52–56
Flow rate, 145
Flow tunnels, 139–141
Force per unit area, 254
Forced polytropic compression, 210
Four-stroke cycle, 162t, 165f
Friction losses, in internal combustion engines, 174
Friction work, 103
Frictional mean effective pressure (FMEP), 179, 271
Froude, William, 4, 5f
Fuel droplets, 129–130
Fuel injection, 61–63, 66
    principles of, 68–72
Fuel temperature conditioning unit, 32–33
Fuel-to-air equivalence ratio, 265
Fuels
    additives for regeneration of catalysts, 120
    air/fuel ratio, 265–266
    composition of, 113–114
    diesel, 114
    gas leakage in internal combustion engines, 168–169
    specific fuel consumption, 254
    test bed flow measurement of, 262
    tests for, 77
    biofuels, 159
    diesel versus gasoline combustion, 130–131
    low emission, 159
Full-scale deflection (FSD), 58

Gauge pressure, 45
Gearbox, automatic, oil and coolant flows for, 27t

Global warming, 113
Glossary, 273–282
Glow plug, 68, 69, 70f
Group A gases, 111
Group B and C gases, 112
Group D gases, 113

Hale-Hamilton valve, 24
Handing over, 234
Hard plugs, 105
Health and Safety at Work Regulations, 1
Heat, 255
Heat engine, 266
Heat exchangers, 24–28
    breakdown symptoms for, 27–28
    header tank, 28–29
Heat losses, in internal combustion engines, 171–174
Heat release, 182–185
Heinrich, Hans, 63
Hesselman, Knut, 61, 62
HMS Conquest, dynamometer for, 4, 5f, 6f, 7f
Hook joint shafts, 19
Hot and cold sampling, 144
Hot junction, 43
    protection of, 44
Hot plugs, 105
Hot spots, 42
Humidity, 31–32, 145
Hydrocarbon traps, 119
Hydrocarbons (HC), 112, 115, 117
Hydrogen, 113–114

IBDC, 210
Idle stability, 185–186
Ignition, compression, 61
Ignition stroke, 162t, 163, 164f
Ignition timing, cold start, 107
Inclined tube manometers, 48
Indicated efficiency, 179
Indicated mean effective pressure (IMEP), 104, 172, 177–178
    quantifying, 188
    root mean square (RMS) of ΔIMEP, 189
Indicated work, 103, 177, 178f
Indirect injection (IDI), 66, 246
    versus direct injection, 131–137
Injection timing, 132, 133f, 136–137
Installation records, 97
Instrumentation
    calibration, 56–59
    flow, 52–56
    pressure, 44–52
    temperature, 41–44
Insulated junction, 44
Intake stroke, 162, 163f,
Integral response, 39

Integral windup, 39
Internal combustion engines, 137–138
    basic, 237–248
    cooling system for, 241
    description of, 162–166
    efficiency loss mechanisms in, 168–170
    emissions from, 125–126
    exhaust system for, 248
    four-stroke cycle in, 162t
    fuel leakage in, 168–169
    heat losses in, 171–174
    induction system for, 245–247
    lubrication system for, 238–241
    noise in, 161–162
    quality assurance for testing of, 249–252
Iridium, 107, 108

Joins, 41
Joule's law, 257–258

Knock, 196–198

Leak checking, 145
Limits of error (LOE), 58
Lloyd's Register Quality Assurance (LRQA), 250
Load, effect of, 131–132
Loop cycle, 39
Lowest normalized value (LNV) of IMEP, 188
Lubricant tests, 77
Lucas-Dawe air mass flow meters, 56

Main oil gallery, 240
Maintenance, of analyzers, 151
Manifold, 246–247
Manometers, 46–48
Mass airflow sensors, 53
Master service record sheet, 94
Maximum heat input, 79
Maximum mechanical and dynamic load, 79
MBT (minimum spark for best torque), 172
McKechnie, James, 61, 62
Mechanical efficiency, 179, 264–265
Mechanical failure, 78
Mechanical power, 256
Metering pumps, 62, 63
Methane, 113–114, 124, 127–128
Methylenecyclopropene (CH_2), 114
Micro-manometers, 47–48
Micro-tunnels, 142–143
Mini-tunnels, 143
Misfire, causes of, 194–195
Motoring mean effective pressure (MMEP), 253, 272

NAMAS Accreditation Standard and Regulations, 250
National Measurement Accreditation Service (NAMAS), 250
Net mean effective pressure (NMEP), 177, 178–179
NGK spark plugs, 107, 109
Nitrogen oxides (NOx), 112, 115, 116
    analyzers for, 148
    efficiency check for, 151
    trap for, 73, 74
Non-dispersive infrared (NDIR) analyzer, 145–146
Non-heat release, 180, 181f
Nonlinear systems, 39

Oil filter, 239, 240
Oil filter bypass valve, 239, 240
Oil flow rate, 241
Oil pressure, 241
Oil pump, 238, 239
Oil temperature, 241
Operational range, 57
Orbital Corporation Ltd., 62
Otto, Nikolaus August, 166
Otto cycle, 237–238
Oxidation efficiency, 217, 218
Oxygen analyzers, 148–149
Oyster curve, 226

Paper element oil filter, 240
Paramagnetic O_2 analyzers, 148–149
Parameter calibration range, 58
Parameter/measuring chain, 57
Partial burn, causes of, 195, 196f
Partial vapor pressure, 261
Particulate analyzers, continuous, 143–144
Particulate measurement, 139
Particulates, new measuring methods for, 158
Pass/fail checklist, 86
Pegging, 209–210
Peltier, Jean, 41
Peltier-Seebeck effect, 41
Percentage overshoot, 38
Performance losses, 95t
Performance tests, 77
Phase, 255
Phasing efficiency losses, in internal combustion engines, 172–174
Photochemical smog, 112
Pickup pipe, 238, 239
Piezoelectric pressure transducers, 50, 51f, 209
Pintle valve, 63, 64f
Pipe-work, 243–244
Piston, damaged, 234–235
Platinum, 107

Plunger system, 65
Polarity, checking, 42, 43f
Polynuclear aromatic hydrocarbons (PAH), 116
Potentiometric pressure transducers, 49
Power, and brake torque, 271
Power correction factor, 268–272
Power stroke, 164
Pre-chamber systems, 66, 68–69, 72
Predictive analysis, 36
Pre-ignition, 103, 104, 106–107, 197, 199
Pressure, 254
    absolute, 45
    air, 30–31
    combustion, 65–66
    control of, 144
    devices for measuring, 45–48
    dry barometric, 261
    gauge, 45
    instrumentation for, 44–52
    oil, 241
    partial, Dalton's law of, 128–129
    partial vapor, 261
    saturated vapor, 261
Pressure losses, in internal combustion engines, 169
Pressure ratio, 231
Pressure relief cap, 243
Pressure relief valve, 238, 239
Pressure transducers, piezoelectric, 50, 51f
Pressure volume, of four-stroke cycle, 165f, 166–168
Pre-start, 36
Process enclosure, 32–33, 34f
Process, 255
Process variable, 38
Prop shafts
    sheared, 235
    selection of, 19–20
Proportional integral derivative (PID), 37
Proportional response, 39
Proportional, integral, and differential gain (PIDG), 38
Psia versus psig, 254–255
Peugeot, 121
Pulse energy, 221–222
Pulse turbocharging, 221–222
Pumping losses, in internal combustion engines, 169, 170f
Pumping mean effective pressure (PMEP), 177, 178, 270

Quality standards, 233, 234, 249–252
    implications of, 251–252

Radiator, 243
Radio frequency interference (RFI), 108
Rassweiler-Withrow algorithm, 182

Raw sampling, 138–139
Rebuilding, 234
Recordkeeping, 96–98, 251
Redundant measures, 215
Reference junction, 43
Reliability, 79, 80
Reporting to customer, 98, 99
Resistor-type spark plugs, 108
Ricardo Consulting Engineers, 176
Rise time, 38
Robustness, 39
Root mean square (RMS) of ΔIMEP, 189
Rotameters, 54

Safety
    with coolant, 244
    with induction system components, 247
    with oil, 241
Sales and marketing, 97, 98
Sample pump, 145
Sampling
    dilute, 138–139
    essential elements for system, 144–151
    raw, 138–139
Saturated vapor pressure, 261
Saunders valves, 28–29
Seebeck, Thomas Johann, 41
Set point, 38
Setting time, 38
Sheathing, 44
Shift change, 234
Siemens Process Automation, 158
Site water services, 22–29
Smog, 112, 115
Smoke meters, 151–152
    filter paper, 153
    opacity, 153–155
Soft plugs, 105
Solid particles, 145
Soluble organic fraction (SOF), 116, 118
Soot, 66, 120
Span gas, 150
Spark plugs, 101–109
    radio frequency interference from, 108
    ratings of, 103–104
    running a rating test on, 104–108
Spark timing, 172
Specific enthalpy, 256
Specific fuel consumption (SFC), 254, 264
Square edged orifice plates, 54–55
Standard deviation of IMEP, 188
Steady-state error, 38, 39
Stewart, Charles, 61
Stoichiometric air/fuel ratio, 66
Strainer, 238, 239
Stripping, 233
Sulfates, 118
Sulfur, 114, 116

Sulfur dioxide (SO_2), 112, 116
Sump, 238, 239
    split, 22–23
Supercharging, 103
Swept volume, 270

Tailpipe emissions, 116–117
Tapered element oscillating microbalance (TEOM), 143–144
Temperature
    air, 30
    calibration of, 58
    control of, 144
    instrumentation for, 41–44
    interior, law of, 42–44
    oil, 241
Terminus, 41
Test bed airflow measurement, 262
Test bed fuel flow measurement, 262
Test cells, 1–4
    air services for, 29–33
    best practices for fitting operations in, 233–235
    calculations required within, 262–263
    cleaning of, 234
    daily checklist for, 95–96
    heat sources in, 32
    in-cell testing, 81, 82
    preventive maintenance for, 95–96
    raw water services for, 21–29
Test cycle, 82
    U.S., 157f
Test laboratories, quality standards for, 250
Tests
    bench, 84–85
    catalyst ageing, 78
    commonly occurring incidents during, 234–235
    durability, 77, 78–79, 83, 86
    duration of, 85
    early development phase, 82
    engine, 77–99
        pointers for, 36–37
    exhaust system, 78
    field, 84–85
    fuel, 77
    interpretation of results, 95
    lubricant, 77
    performance, 77
    severe, 81, 82–83
    specialized, 78
    thermal shock, 84
    total vehicle certified emission, 156–157
    transient, 37
    types of, 77–78
    validation, typical, 86–93
Thermal efficiency, 236–238, 264
Thermal shock testing, 84

Thermal stress, 84
Thermocouples, 41–44
    categories of, 44
        mineral-insulated metal-sheathed (MIMS), 42, 43f
    specifications for, 45t
Thermodynamic heat release, 184–185
Thermodynamics, laws of, 257, 266–268
Thermoelectric junction, 41
Thermo-element, 41
Thermostat, 242
Thermostat housing, 243
Three-way conversion, 137
Throttle butterfly valve, 246
Time alignment, 150
Time-to-combust losses, in internal combustion engines, 169, 170f
Timing belt, "jumping a tooth," 235
Top dead center (TDC), 162
Torque, 16
    brake, 271
Torque backup, calculation of, 253
Total preventive maintenance (TPM), 251
Total vehicle certified emission test, 156–157
Traceability, 58
Transducers, pressure, 48–52
    mounting considerations for, 209
    piezoelectric, 209
    recommendations for using, 210–211
Transient test, 37
Tuning, 40
Turbine flow meter, 52, 53f
Turbocharger matching, 225–231
Turbochargers, 221–231
Turbocharging
    constant pressure, 221
    pulse, 221–222

Uncertainty of measurement, 58
Unit injectors, 65
United Kingdom Accreditation Service (UKAS), 250
U-tube manometers, 46, 47f

Valve, damaged, 234–235
Variable fill hydraulic dynamometers, 11–12
Variable vane geometry turbocharger (VGT), 229, 230f
Ventilation, 31, 32–33
Venturi gas meter, 52, 53f
Vicars Company, 62
Volumetric efficiency, 179–180, 253–254, 265

Waste gate turbocharger, 228, 229f
Water brakes, 7–12, 17f, 18, 36
Water pump, 242
Water services for test cell, 21–29
Wear rates, 95t
Wet basis calculation, 129
Whirl-chamber system, 68, 69f, 72
Wiedman, Carl, 65
Wika transducers, 51

Zeolites, 119
Zero gas, 150
Zeroth law of thermodynamics, 257

About the Author

Throughout his career, Richard D. Atkins has worked with the internal combustion engine. He was closely involved in the development of Formula 1 race engines through the 1960s, and he was the development manager with Tecalemit, covering the design and development of early electronic fuel injection systems used in race and rally cars. Mr. Atkins was responsible for the build and running of cars in both the London-to-Sydney and the London Mexico World Cup International rallies.

Mr. Atkins formed his own company to manufacture electronic fuel injection systems for race cars and motorcycles. He has worked in that capacity as a consultant and for major automotive companies in France, Italy, Germany, Denmark, Spain, Poland, Slovakia, India, and Iran.

Mr. Atkins has been an active member of SAE International since 1988. He also is a Chartered Engineer and a Fellow of the Institute of Mechanical Engineers, where he holds the post of Chairman of the Automobile Division Southern Centre.

While continuing his consultancy business, Mr. Atkins currently is a Visiting Senior Research Fellow at Sussex University and a Visiting Fellow at Kingston Universities in the United Kingdom, where his extensive knowledge of the internal combustion engine is greatly valued.